Earth Science: A Comprehensive Approach

Earth Science: A Comprehensive Approach

Adam Rush

Larsen & Keller
www.larsen-keller.com

Earth Science: A Comprehensive Approach
Adam Rush
ISBN: 978-1-64172-468-5 (Hardback)

▤ Larsen & Keller

Published by Larsen and Keller Education,
5 Penn Plaza,
19th Floor,
New York, NY 10001, USA

Cataloging-in-Publication Data

Earth science : a comprehensive approach / Adam Rush.
    p. cm.
Includes bibliographical references and index.
ISBN 978-1-64172-468-5
1. Earth sciences. 2. Physical sciences. 3. Geology. 4. Geography. I. Rush, Adam.
QE26.3 .E27 2020
550--dc23

For more information regarding Larsen and Keller Education and its products, please visit the publisher's website www.larsen-keller.com

Table of Contents

Preface

The set of natural sciences which focuses on the planet Earth are known as earth science. It primarily deals with studying the physical characteristics of Earth. There are four major areas of study within this field. These include the atmosphere, the hydrosphere, the lithosphere and the biosphere. Each of these is further sub-divided into more specialized fields. Earth science makes use of tools from a variety disciplines such as chronology, geography, chemistry, biology, physics and mathematics. These are used to construct a quantitative understanding of the working and evolution of the Earth. This discipline is also involved in the study of different geological phenomena related to the Earth's crust, mantle and core. This book provides significant information of this discipline to help develop a good understanding of earth science. Those with an interest in this field would find this book helpful. It is appropriate for students seeking detailed information in this area as well as for experts.

A foreword of all Chapters of the book is provided below:

Chapter 1 - The field of natural science which is involved in studies related to the planet Earth is known as Earth science. Some of the major branches of Earth science are geology, geography, oceanography, meteorology and astronomy. This chapter will provide a brief introduction of these branches of Earth science.; **Chapter 2** - The internal structure of the Earth is made up of several layers, namely, the crust, the mantle and the core. Crust is further sub-divided into oceanic crust and continental crust. The topics elaborated in this chapter will help in gaining a better perspective about these components of the Earth's structure as well as their properties.; **Chapter 3** - Earth's spheres refer to the subsystems which make up the natural environment of the Earth. The four major spheres of the Earth are atmosphere, biosphere, hydrosphere and the geosphere. The chapter closely examines these key spheres of the Earth to provide an extensive understanding of the subject.; **Chapter 4** - There are numerous cycles which are involved in regulating and balancing the Earth and its atmosphere. A few of them are oxygen cycle, carbon cycle, nitrogen cycle, sulfur cycle and phosphorus cycle. This chapter has been carefully written to provide an easy understanding of the varied facets of these Earth's cycles.; **Chapter 5** - An ecosystem consists of all the living things in an area such as plants, animals and other organisms. The major types of ecosystems are aquatic ecosystems and terrestrial ecosystems. This chapter closely examines these primary ecosystems to provide an extensive understanding of the subject.; **Chapter 6** - There are varied elements which are found on the Earth such as minerals, rocks, soil and water. The minerals are classified into numerous categories such as nonmetals, metals and semimetals. This chapter has been carefully written to provide an easy understanding of the properties and types of these elements.

I would like to thank the entire editorial team who made sincere efforts for this book and my family who supported me in my efforts of working on this book. I take this opportunity to thank all those who have been a guiding force throughout my life.

Adam Rush

An Introduction to Earth and Earth Science

The field of natural science which is involved in studies related to the planet Earth is known as Earth science. Some of the major branches of Earth science are geology, geography, oceanography, meteorology and astronomy. This chapter will provide a brief introduction of these branches of Earth science.

Earth

Earth is the third planet from the Sun and the fifth largest planet in the solar system in terms of size and mass. Its single most outstanding feature is that its near-surface environments are the only places in the universe known to harbour life. It is designated by the symbol ♁. Earth's name in English, the international language of astronomy, derives from Old English and Germanic words for ground and earth, and it is the only name for a planet of the solar system that does not come from Greco-Roman mythology.

Earth: A composite image of Earth captured by instruments aboard NASA's Suomi National Polar-orbiting Partnership satellite.

Since the Copernican revolution of the 16th century, at which time the Polish astronomer Nicolaus Copernicus proposed a Sun-centred model of the universe; enlightened thinkers have regarded Earth as a planet like the others of the solar system. Concurrent sea voyages provided practical proof that Earth is a globe, just as Galileo's use of his newly invented telescope in the early 17th century soon showed various other planets to be globes as well. It was only after the dawn of the space age, however, when photographs from rockets and orbiting spacecraft first captured the dramatic curvature of Earth's horizon, that the conception of Earth as a roughly spherical planet rather than as a flat entity was verified by direct human observation. Humans first witnessed Earth as a complete orb floating in the inky blackness of space in December 1968 when Apollo 8 carried astronauts around the Moon. Robotic space probes on their way to destinations beyond Earth, such as the Galileo and the Near Earth Asteroid Rendezvous (NEAR) spacecraft in the 1990s, also looked back with their cameras to provide other unique portraits of the planet.

Viewed from another planet in the solar system, Earth would appear bright and bluish in colour. Easiest to see through a large telescope would be its atmospheric features, chiefly the swirling white cloud patterns of multitude and tropical storms, ranged in roughly latitudinal belts around the planet. The polar region also would appear a brilliant white, because of the clouds above and the snow and ice below. Beneath the changing patterns of clouds would appear the much darker blue-black oceans, interrupted by occasional tawny patches of desert lands. The green landscapes that harbour most human life would not be easily seen from space. Not only do they constitute a modest fraction of the land area, which itself is less than one-third of Earth's surface, but they are often obscured by clouds. Over the course of the seasons, some changes in the storm patterns and cloud belts on Earth would be observed. Also prominent would be the growth and recession of the winter snow-cap across land areas of the Northern Hemisphere.

Scientists have applied the full battery of modern instrumentation to studying Earth in ways that have not yet been possible for the other planets; thus, much more is known about its structure and composition. This detailed knowledge, in turn, provides deeper insight into the mechanisms by which planets in general cool down, by which their magnetic fields are generated, and by which the separation of lighter elements from heavier ones as planets develop their internal structure releases additional energy for geologic processes and alters crustal compositions.

Earth's surface is traditionally subdivided into seven continental masses: Africa, Antarctica, Asia, Australia, Europe, North America, and South America. These continents are surrounded by four major bodies of water: the Arctic, Atlantic, Indian, and Pacific oceans. However, it is convenient to consider separate parts of Earth in terms of concentric, roughly spherical layers. Extending from the interior outward, these are the core, the mantle, the crust (including the rocky surface), the hydrosphere (predominantly the oceans, which fill in low places in the crust), the atmosphere (itself divided into spherical zones such as the troposphere, where weather occurs, and the stratosphere, where lies the ozone layer that shields Earth's surface and its organisms against the Sun's ultraviolet rays), and the magnetosphere (an enormous region in space where Earth's magnetic field dominates the behaviour of electrically charged particles coming from the Sun).

Earth Science

Earth Science is the study of the Earth and its neighbors in space. It is an exciting science with many interesting and practical applications. Some Earth scientists use their knowledge of the Earth to locate and develop energy and mineral resources. Others study the impact of human activity on Earth's environment, and design methods to protect the planet. Some use their knowledge about Earth processes such as volcanoes, earthquakes, and hurricanes to plan communities that will not expose people to these dangerous events.

Branches of Earth Science

Geology

Geology is an earth science concerned with the solid Earth, the rocks of which it is composed, and the processes by which they change over time. Geology can also refer generally to the study of the solid features of any terrestrial planet (such as the geology of the Moon or Mars).

Geology gives insight into the history of the Earth by providing the primary evidence for plate tectonics, the evolutionary history of life, and past climates. Geology is important for mineral and hydrocarbon exploration and exploitation, evaluating water resources, understanding of natural hazards, the remediation of environmental problems, and for providing insights into past climate change. Geology also plays a role in geotechnical engineering and is a major academic discipline.

Important Principles of Geology

There are a number of important principles in geology. Many of these involve the ability to provide the relative ages of strata or the manner in which they were formed.

- The Principle of Intrusive Relationships concerns crosscutting intrusions. In geology, when an igneous intrusion cuts across a formation of sedimentary rock, it can be determined that the igneous intrusion is younger than the sedimentary rock. There are a number of different types of intrusions, including stocks, laccoliths, batholiths, sills, and dikes.

- The Principle of Cross-cutting Relationships pertains to the formation of faults and the age of the sequences through which they cut. Faults are younger than the rocks they cut; accordingly, if a fault is found that penetrates some formations but not those on top of it, then the formations that were cut are older than the fault, and the ones that are not cut must be younger than the fault. Finding the key bed in these situations may help determine whether the fault is a normal fault or a thrust fault.

- The Principle of Inclusions and Components states that with sedimentary rocks, if inclusions (or *clasts*) are found in a formation, then the inclusions must be older than the formation that contains them. For example, in sedimentary rocks, it is common for gravel from an older formation to be ripped up and included in a newer layer. A similar situation with igneous rocks occurs when xenoliths are found. These foreign bodies are picked up as magma or lava flows, and are incorporated later to cool in the matrix. As a result, xenoliths are older than the rock which contains them.

- The Principle of Uniformitarianism states that the geologic processes observed in operation that modify the Earth's crust at present have worked in much the same way over geologic time. A fundamental principle of geology advanced by the eighteenth-century Scottish physician and geologist James Hutton is that "The Present is the Key to the Past." In Hutton's words: "the past history of our globe must be explained by what can be seen to be happening now."

- The Principle of Original Horizontality states the deposition of sediments occurs as essentially horizontal beds. Observation of modern marine and non-marine sediments in a wide variety of environments supports this generalization (although cross-bedding is inclined, the overall orientation of cross-bedded units is horizontal).

- The Principle of Superposition states a sedimentary rock layer in a tectonically undisturbed sequence is younger than the one beneath it and older than the one above it. Logically a younger layer cannot slip beneath a layer previously deposited. This principle allows sedimentary layers to be viewed as a form of vertical time line, a partial or complete record of the time elapsed from deposition of the lowest layer to deposition of the highest bed.

- The Principle of Faunal Succession is based on the appearance of fossils in sedimentary rocks. As organisms exist at the same time period throughout the world, their presence or (sometimes) absence may be used to provide a relative age of the formations in which they are found. Based on principles laid out by William Smith almost a hundred years before the publication of Charles Darwin's theory of evolution, the principles of succession were developed independently of evolutionary thought. The principle becomes quite complex, however, given the uncertainties of fossilization, the localization of fossil types due to lateral changes in habitat (fancies change in sedimentary strata), and that not all fossils may be found globally at the same time.

Some Branches of Geology

Geochemistry

Geochemistry is the study of the chemical processes which form and shape the Earth. It includes the study of the cycles of matter and energy which transport the Earth's chemical components and the interaction of these cycles with the hydrosphere and the atmosphere.

It is a subfield of inorganic chemistry, which is concerned with the properties of all the elements in the periodic table and their compounds. Inorganic chemistry investigates the characteristics of substances that are not organic, such as non-living matter and minerals found in the Earth's crust.

Paleontology

Paleontologists are interested in fossils and how ancient organisms lived. Paleontology is the study of fossils and what they reveal about the history of our planet. In marine environments, microfossils collected within in layers of sediment cores provide a rich source of information about the environmental history of an area.

Sedimentology

Sedimentology is the study of sediment grains in marine and other deposits, with a focus on physical properties and the processes which form a deposit. Deposition is a geological process where geological material is added to a landform. Key physical properties of interest include:

- The size and shape of sediment grains;
- The degree of sorting of a deposit;
- The composition of grains within a deposit;
- Sedimentary structures.

These properties together provide a record of the mechanisms active during sediment transportation and deposition which allows the interpretation of the environmental conditions that produced a sediment deposit, either in modern settings or in the geological record.

Additional Branches

- Benthic Ecology: Benthic ecology is the study of living things on the seafloor and how they interact with their environment.
- Biostratigraphy: Biostratigraphy is the branch of stratigraphy that uses fossils to establish relative ages of rock and correlate successions of sedimentary rocks within and between depositional basins.

- Geochronology: Geochronology is a discipline of geoscience which measures the age of earth materials and provides the temporal framework in which other geoscience data can be interpreted in the context of Earth history.

- Geophysics: Information relating to various techniques including: airborne electromagnetics, gravity, magnetics, magnetotellurics, radiometrics, rock properties and seismic.

- Marine Geochemistry. Marine geochemistry is the science used to help develop an understanding of the composition of coastal and marine water and sediments.

- Marine Geophysics: Marine geophysics is a scientific discipline which uses the quantitative observation of physical properties to understand the seafloor and sub-seafloor geology.

- Marine Surveying: The survey environment varies from oceanographic studies in the water column to investigating sediment and geochemical processes on the seafloor and imaging the sub-seafloor rocks. Surveys are carried over Australia's entire marine jurisdiction, from coastal estuaries and bays, across the continental shelf and slope, to the deep abyssal plains.

- Spectral Geology: Spectral geology is the measurement and analysis of portions of the electromagnetic spectrum to identify spectrally distinct and physically significant features of different rock types and surface materials, their mineralogy and their alteration signatures.

Geography

Geography is the study of life on the surface of the earth. Geography is fundamentally interdisciplinary. It is one of the few disciplines that encompass very different ways of knowing, from the natural and social sciences and the humanities.

The Branches of Geography

The subject encompasses an interdisciplinary perspective that allows the observation and analysis of anything distributed in Earth space and the development of solutions to problems based on such analysis. The discipline of geography can be divided into several branches of study. The primary classification of geography divides the approach to the subject into the two broad categories of physical geography and human geography.

Physical Geography

Physical geography is defined as the branch of geography that encompasses the study of the natural features and phenomena (or processes) on the Earth.

Physical geography may be further subdivided into various branches:

1. Geomorphology: This involves the study of the topographic and bathymetric features on Earth. The science helps to elucidate various aspects related to the landforms on the Earth such as their history and dynamics. Geomorphology also attempts to predict the future changes in the Earth's physical features.

2. Glaciology: This field of physical geography deals with the study of the inter-dynamics of glaciers and their effects on the planet's environment. Thus, glaciology involves the study of the cry sphere including the alpine glaciers and the continental glaciers. Glacial geology, snow hydrology, etc., are some of the sub-fields of glaciological studies.

3. Oceanography: Since oceans hold 96.5% of the Earth's waters, a special field of oceanography needs to be dedicated to the study of oceans. The science of oceanography includes geological oceanography (study of the geological aspects of the ocean floor, its mountains, volcanoes, etc.), biological oceanography (study of the marine life and ocean ecosystems), chemical oceanography (study of the chemical composition of the marine waters and their effects on marine life forms), physical oceanography (study of the oceanic movements like the waves, currents, etc.)

4. Hydrology: This is another vital aspect of physical geography. Hydrology deals with the study of the properties of the Earth's water resources and the movement dynamics of water in relation to land. The field encompasses the study of the rivers, lakes, glaciers, and underground aquifers on the planet. It studies the continuous movement of water from one source to another on, above, and below the Earth's surface, in the form of the hydrological cycle.

5. Pedology: A branch of soil science, pedology involves the study of the different soil types in their natural environment on the surface of the Earth. This field of study helps gather information and knowledge on the process of soil formation (pedogenesis), soil constitution, soil texture, classification, etc.

6. Biogeography: An indispensable field of physical geography, biogeography is the study of how species on Earth are dispersed in geographic space. It also deals with the distribution of species over geological time periods. Each geographical area has its own unique ecosystem and biogeography explores and explains such ecosystems in relation to physical geographical features. Different branches of biogeography exist like zoogeography (geographic distribution of animals), phytogeography (geographic distribution of plants), insular biogeography (the study of factors influencing isolated ecosystems), etc.

7. Paleogeography: This branch of physical geography examines the geographical features at various time points in the Earth's geological history. It helps the

geographers to attain knowledge about the continental positions and plate tectonics determined by studying paleomagnetism and fossil records.

8. Climatology: The scientific study of climate, climatology is a crucial field of geographical studies in today's world. It examines all aspects related to the micro or local climates of places and also the macro or global climate. It also involves an examination of the impact of human society on climate and vice versa.

9. Meteorology: This field of physical geography is concerned with the study of the weather patterns of a place and the atmospheric processes and phenomena that influence the weather.

10. Environmental geography: Also known as integrative geography, this field of physical geography explores the interactions between humans (individuals or society) and their natural environment from the spatial point of view. Environmental geography is thus the bridging gap between human geography and physical geography and can be treated as an amalgamation of multiple fields of physical geography and human geography.

11. Coastal geography: Coastal geography is another area of specialization of physical geography that also involves a study of human geography. It deals with the study of the dynamic interface between the coastal land and the sea. The physical processes that shape the coastal landscape and the influence of the sea in triggering landscape modifications is incorporated in the study of coastal geography. The study also involves an understanding of the ways the human inhabitants of coastal areas influence the coastal landforms and ecosystems.

12. Quaternary science: This is a highly specialized field of physical geography that deals with the study of the Quaternary period on Earth (the Earth's geographical history encompassing the last 2.6 million years). It allows the geographers to learn about the environmental changes undergone in the planet's recent past. This knowledge is then used as a tool to predict future changes in the Earth's environment.

13. Geomatics: Geomatics is a technical branch of physical geography that involves the collection of data related to the earth's surface, analysis of the data, its interpretation, and storage. Geodesy, remote sensing, and geographical information science are the three sub-divisions of geomatics.

14. Landscape ecology: The science of landscape ecology deals with the study of how the varying landscapes on Earth influences the ecological processes and ecosystems on the planet. The German geographer Carl Troll is credited as the founder of this field of physical geography.

Human Geography

Human geography is the branch of geography that deals with the study of how the human society is influenced by the Earth's surface and environment and how, in turn, anthropological activities impact the planet. Human geography is centered on the study of the planet's most evolved creatures: the humans and their environment.

This branch of geography can be further subdivided into various disciplines based on the focus of study:

1. Population geography: A division of human geography, population geography deals with the study of how the nature of a place determines the distribution, growth, composition, and migration of human populations.

2. Historical geography: Historical geography elucidates the ways in which geographical phenomena change and evolve with time. Though it is treated as a subfield of human geography, it also focuses on certain aspects of physical geography. Historical geography attempts to understand why, how and when a place or region on Earth changes and the impact such changes have on the human society.

3. Cultural geography: Cultural geography explores how and why cultural products and norms vary with space and place. It thus deals with the study of the spatial variations of human cultures including religion, language, livelihood choices, politics, etc. Religion geography, language geography, etc., are some of the subfields of cultural geography.

4. Economic geography: A vital aspect of human geography, economic geography encompasses the study of how human economic activities are located, distributed and organized in geographical place and space. Marketing and transportation geography can be treated as sub-fields of economic geography.

5. Political geography: This important field of human geography deals with the political boundaries of the countries of the world and the division of land and its resources between the countries. It also deals with how spatial structures influence political functions and vice versa. Military geography, electoral geography, geopolitics are some of the subfields of political geography.

6. Health geography: A sub-discipline of human geography, health geography concentrates on the influence of the geographical location and place on the health and well-being of humans. It tends to approach the subject of human health from a comprehensive perspective encompassing the influence of society and space on health and disease.

7. Developmental geography: This branch of human geography explores the quality of life and the standard of living of the human inhabitants of the world and attempts to understand how and why such standards vary with place and space.

8. Settlement geography: Settlement geography attempts to explore the part of the Earth's surface that encompasses human settlements. It is a study of the urban and rural settlements, the economic structure, infrastructure, etc., and the dynamics of human settlement patterns in relation to space and time.

9. Animal geography: Animal geography might be considered as a sub-field of human geography which is closely related to the environmental geography branch of physical geography. It encompasses the study of the life worlds of the animals on Earth and the interdependencies between humans and other animals.

Oceanography

Oceanography is the scientific study of oceans and seas. It deals with the distribution of oceanic water masses, morphology and relief of the ocean floors, depth zones in oceans, sediments of the oceans, marine mineral resources, oceanic processes dynamics of water masses and the role of oceans on controlling the global climate. The physics, chemistry, geology and biology of the oceans are very deep concepts in natural sciences. Oceans are very dynamic and widely distributed water bodies. The nature and role of oceans are continuously studied by the oceanographers, marine biologists, marine engineers, environmental scientists, ecologists, geologists, marine geologists, meteorologists, climatologists and geographers. It is one of the oldest subjects of mankind. Even today, the subject of oceanography is diversifying into many folds and branches.

Branches of Oceanography

The major branches of oceanography are:

1. Physical oceanography,

2. Chemical oceanography,

3. Geological oceanography,

4. Biological oceanography.

Physical Oceanography

The Physical Oceanography is an essential part of oceanographic analysis. It is the study of physical conditions that are prevailing in the seas and oceans. It deals with all large scale physical processes and their effects that are happening within the oceans. the processes are very dynamic in nature. They involve the water masses which have heterogeneous proportions and dimension. The properties of water masses also vary with space and time. Physical oceanography considers all these aspects in projecting the oceans. This branch has grown due to historical oceanographic explorations and expeditions carried out by several scholars from almost all parts of the world. Very essential aspects of physical oceanography are to be understood first.

Chemical Oceanography

Chemical Oceanography is the study of everything about the chemistry of the ocean, distribution and dynamics of the elements, isotopes, atoms and molecules. This ranges from fundamental physical, thermodynamic and kinetic chemistry to two-way interactions of ocean chemistry with biological/ geological and physical processes. Chemical Oceanography attempts to analyse the interactions between oceans, lithosphere, atmosphere and biosphere, sea water chemistry, controls in chemical distribution, components of marine sediments and chemical controls in biological production.

Biological Oceanography

The basic ecological concepts are central to many studies of biological oceanography. The study of marine life, habitat, interactions, abiotic environment, phytoplankton and primary production, zooplankton, migrations and changes, energy flow & mineral cycling, marine food chains, food webs, nektons, marine reptiles, mammals, seabirds, mariculture, Benthic plants and animals, inter-tidal environments, beaches, coral reefs, estuaries and mangroves are all studied under biological oceanography. Deep sea ecology and marine pollution are also the other two major important areas of study under biological oceanography.

Geological Oceanography

Geological Oceanography is a division of oceanography mainly dealing with the basic Concepts of lithosphere & hydrosphere. It includes the study of the oceanic crust, continental margins, ocean bottom relief, ocean basins, oceanic ridges, rift-valleys, Island arcs, sea water, marine sedimentation, geology of corals, beach forms and processes, water masses , factors affecting ocean circulation, waves and currents, tides and energy coastal erosion and drifting of sediments, sea level changes, depositional environments and marine deposits. Geological oceanography is also concerned with the occurrence of oil-traps and energy sources, tectonic movements- underwater eruptions, mud volcanoes and impacts of tsunamis.

Meteorology

Meteorology is a scientific study that focuses on the atmosphere and weather processes including forecasting. Observable weather events, also known as meteorological phenomena, are included in the study of meteorology. Conventional meteorological studies include water vapor, temperature, and air pressure as well as the gradients of each and how each of these variables interacts with one another. Meteorology also studies how these variables change throughout time. Most of the Earth's weather occurs within the troposphere.

Types of Meteorology

Boundary layer meteorology studies the layer of air that is directly above the Earth's surface. It focuses on how the surface-atmosphere boundary affects the oceans, lakes, urban land areas and non-urban land areas. This study delves into how the heating, cooling, and friction affects these areas and how these cause turbulent mixing in the boundary layer of air.

Mesoscale meteorology studies several layers of the Earth including the boundary layer, the stratosphere, the tropopause and the troposphere. It examines several forms of weather such as land breezes, mountain waves, thunderstorms, fronts, squall lines and precipitation bands in extra-tropical and tropical cyclones. These weather events are then often followed throughout their duration.

Global scale meteorology studies the weather patterns that are significant to the transport of heat from the north and south poles to the tropics. It also focuses on oscillations of a vast scale. It delves into how global oscillations cause weather disturbances and climate disturbances in mesoscale timescales and synoptic timescales.

Agricultural meteorology studies how weather affects plant development, animal development, plant distribution, crop yield and the efficiency of water use. This type of meteorology studies the energy balance of natural ecosystems and managed ecosystems for similarities and differences. Those in this field are also looking into how vegetation may affect the weather and climate.

Hydrometeorology studies the hydrologic cycle tracks the storms' rainfall statistics and deals with the water budget. This type of meteorologist prepares the forecasts that are often in the news and gives the information on amounts of potential precipitation, snow, and rain and predicts areas that may be in danger of flash floods.

Importance of Meterology

Meteorology plays a significant role in environmental science. It is helpful for determining and tracking climate patterns as well as how land and water play a part in the climate and climate change. It gives information on oscillations and how global oscillations may cause weather and climate disturbances. Meteorology also helps to control vegetation, and it also allows those in agriculture determine the best time to promote their crops. Meteorology also studies how past climates can predict future climates and the dangers that certain environmental processes have on Earth.

Astronomy

Astronomy is science that encompasses the study of all extra-terrestrial objects and phenomena. Until the invention of the telescope and the discovery of the laws of motion and gravity in the 17th century, astronomy was primarily concerned with noting and predicting the positions of the Sun, Moon, and planets, originally for calendrical and astrological purposes and later for navigational uses and scientific interest. The catalog of objects now studied is much broader and includes, in order of increasing distance, the solar system, the stars that make up the Milky Way Galaxy, and other, more distant galaxies. With the advent of scientific space probes, Earth also has come to be studied as one of the planets, though its more-detailed investigation remains the domain of the Earth sciences.

Techniques of Astronomy

Astronomical observations involve a sequence of stages, each of which may impose constraints on the type of information attainable. Radiant energy is collected with tele-scopes and brought to a focus on a detector, which is calibrated so that its sensitivity and spectral response are known. Accurate pointing and timing are required to permit the correlation of observations made with different instrument systems working in different wavelength intervals and located at places far apart. The radiation must be spectrally analyzed so that the processes responsible for radiation emission can be identified.

Telescopic Observations

Before Galileo Galilei's use of telescopes for astronomy in 1609, all observations were made by naked eye, with corresponding limits on the faintness and degree of detail that could be seen. Since that time, telescopes have become central to astronomy. Having apertures much larger than the pupil of the human eye, telescopes permit the study of faint and distant objects. In addition, sufficient radiant energy can be collected in short time intervals to permit rapid fluctuations in intensity to be detected. Further, with more energy collected, a spectrum can be greatly dispersed and examined in much greater detail.

Aerial view of the Keck Observatory's twin domes, which are opened to reveal the telescopes. Keck II is on the left and Keck I on the right.

Optical telescopes are either refractors or reflectors that use lenses or mirrors, respectively, for their main light-collecting elements (objectives). Refractors are effective-ly limited to apertures of about 100 cm (approximately 40 inches) or less because of problems inherent in the use of large glass lenses. These distort under their own weight and can be supported only around the perimeter; an appreciable amount of light is lost due to absorption in the glass. Large-aperture refractors are very long and require large and expensive domes. The largest modern telescopes are all reflectors, the very largest composed of many segmented components and having overall diameters of about 10 metres (33 feet). Reflectors are not subject to the chromatic problems of refractors, can be better supported mechanically, and can be housed in smaller domes because they are more compact than the long-tube refractors.

The angular resolving power (or resolution) of a telescope is the smallest angle between close objects that can be seen clearly to be separate. Resolution is limited by the wave nature of light. For a telescope having an objective lens or mirror with diameter D and operating at wavelength λ, the angular resolution (in radians) can be approximately described by the ratio λ/D. Optical telescopes can have very high intrinsic resolving powers; in practice, however, these are not attained for telescopes located on Earth's surface, because atmospheric effects limit the practical resolution to about one arc second. Sophisticated computing programs can allow much-improved resolution, and the performance of telescopes on Earth can be improved through the use of adaptive optics, in which the surface of the mirror is adjusted rapidly to compensate for atmospheric turbulence that would otherwise distort the image. In addition, image data from several telescopes focused on the same object can be merged optically and through computer processing to produce images having angular resolutions much greater than that from any single component.

The atmosphere does not transmit radiation of all wavelengths equally well. This restricts astronomy on Earth's surface to the near ultraviolet, visible, and radio regions of the electromagnetic spectrum and to some relatively narrow "windows" in the nearer infrared. Longer infrared wavelengths are strongly absorbed by atmospheric water vapour and carbon dioxide. Atmospheric effects can be reduced by careful site selection and by carrying out observations at high altitudes. Most major optical observatories are located on high mountains, well away from cities and their reflected lights. Infrared telescopes have been located atop Mauna Kea in Hawaii, in the Atacama Desert in Chile, and in the Canary Islands, where atmospheric humidity is very low. Airborne telescopes designed mainly for infrared observations—such as on the Stratospheric Observatory for Infrared Astronomy (SOFIA), a jet aircraft fitted with astronomical instruments—operate at an altitude of about 12 km (40,000 feet) with flight durations limited to a few hours. Telescopes for infrared, X-ray, and gamma-ray observations have been carried to altitudes of more than 30 km (100,000 feet) by balloons. Higher altitudes can be attained during short-duration rocket flights for ultraviolet observations. Telescopes for all wavelengths from infrared to gamma rays have been carried by robotic spacecraft observatories such as the Hubble Space Telescope and the Wilkinson Microwave Anisotropy Probe, while cosmic rays have been studied from space by the Advanced Composition Explorer.

The James Clerk Maxwell Telescope located near the summit of Mauna Kea.

Angular resolution better than one milliarcsecond has been achieved at radio wavelengths by the use of several radio telescopes in an array. In such an arrangement, the effective aperture then becomes the greatest distance between component telescopes. For example, in the Very Large Array (VLA), operated near Socorro, New Mexico, by the National Radio Astronomy Observatory, 27 movable radio dishes are set out along tracks that extend for nearly 21 km. In another technique, called very long baseline interferometry (VLBI), simultaneous observations are made with radio telescopes thousands of kilometres apart; this technique requires very precise timing.

Very Large Array: Very Large Array, radio telescope system located on the plains of San Agustin, near Socorro, New Mexico.

Earth is a moving platform for astronomical observations. It is important that the specification of precise celestial coordinates be made in ways that correct for telescope location, the position of Earth in its orbit around the Sun, and the epoch of observation, since Earth's axis of rotation moves slowly over the years. Time measurements are now based on atomic clocks rather than on Earth's rotation and telescopes can be driven continuously to compensate for the planet's rotation, so as to permit tracking of a given astronomical object.

Use of Radiation Detectors

Although the human eye remains an important astronomical tool, detectors capable of greater sensitivity and more rapid response are needed to observe at visible wavelengths and, especially, to extend observations beyond that region of the electromagnetic spectrum. Photography was an essential tool from the late 19th century until the 1980s, when it was supplanted by charge-coupled devices (CCDs). However, photography still provides a useful archival record. A photograph of a particular celestial object may include the images of many other objects that were not of interest when the picture was taken but that become the focus of study years later. When quasars were discovered in 1963, for example, photographic plates exposed before 1900 and held in the Harvard College Observatory were examined to trace possible changes in position or intensity of the radio object newly identified as quasar 3C 273. Also, major photographic surveys, such as those of the National Geographic Society and the Palomar Observatory, can provide a historical base for long-term studies.

Photographic film converted only a few percent of the incident photons into images, whereas CCDs have efficiencies of nearly 100 percent. CCDs can be used for a wide range of wavelengths, from the X-ray into the near-infrared. Gamma rays are detectable through their Compton scattering, electron-positron pair production, or Cerenkov radiation. For infrared wavelengths longer than a few microns, semiconductor detectors that operate at very low (cryogenic) temperatures are used. Reception of radio waves is based on the production of a small voltage in an antenna rather than on photon counting.

Spectroscopy involves measuring the intensity of the radiation as a function of wavelength or frequency. In some detectors, such as those for X-rays and gamma rays, the energy of each photon can be measured directly. For low-resolution spectroscopy, broadband filters suffice to select wavelength intervals. Greater resolution can be obtained with prisms, gratings, and interferometers.

Multi-messenger Astronomy

Most of what is known about the universe comes from observations of electromagnetic radiation. However, there are other "cosmic messengers." Gravity waves are disturbances in space-time that can be detected by very large laser interferometers. Gravity waves and gamma-ray bursts have been observed from neutron-star mergers. Neutrinos and cosmic rays are other particles that can, in principle, be observed; however, as yet, these latter messengers cannot be identified with specific sources. Using two or more of these methods is called multi-messenger astronomy.

Laser Interferometer Gravitational-Wave Observatory (LIGO): The Laser Interferometer Gravitational-Wave Observatory (LIGO) near Hanford, Washington, U.S. There are two LIGO installations; the other is near Livingston, Louisiana, U.S.

Solid Cosmic Samples

As a departure from the traditional astronomical approach of remote observing, certain more recent lines of research involve the analysis of actual samples under laboratory conditions. These include studies of meteorites, rock samples returned from the Moon, cometary and asteroid dust samples returned by space probes, and interplanetary dust particles collected by aircraft in the stratosphere or by spacecraft. In all such cases, a wide range of highly sensitive laboratory techniques can be adapted for the often microscopic samples. Chemical analysis can be supplemented with mass spectrometry, allowing isotopic composition to be determined. Radioactivity and the impacts of cosmic-ray particles can produce minute quantities of gas, which then remain trapped in crystals within the samples. Carefully controlled heating of the crystals (or of dust grains containing the crystals) under laboratory conditions releases this gas, which then is analyzed in a mass spectrometer. X-ray spectrometers, electron microscopes, and microprobes are employed to determine crystal structure and composition, from which temperature and pressure conditions at the time of formation can be inferred.

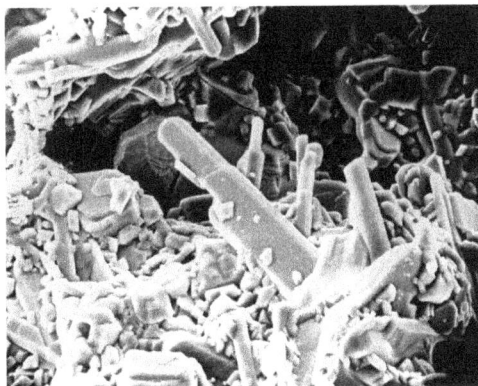

Moon rock; crystals: A scanning-electron-microscope photograph of pyroxene and plagioclase crystals (the long and the short crystals, respectively) that grew in a cavity in a fragment of Moon rock gathered during the Apollo 14 mission.

Theoretical Approaches

Theory is just as important as observation in astronomy. It is required for the interpretation of observational data; for the construction of models of celestial objects and physical processes, their properties, and their changes over time; and for guiding further observations. Theoretical astrophysics is based on laws of physics that have been validated with great precision through controlled experiments. Application of these laws to specific astrophysical problems, however, may yield equations too complex for direct solution. Two general approaches are then available. In the traditional method, a simplified description of the problem is formulated, incorporating only the major physical components, to provide equations that can be either solved directly or used to create a numerical model that can be evaluated. Successively more-complex models can then be investigated. Alternatively, a computer program can be devised that will explore the problem numerically in all its complexity. Computational science has taken its place as a major division alongside theory and experiment. The test of any theory is its ability to incorporate the known facts and to make predictions that can be compared with additional observations.

Impact of Astronomy

No area of science is totally self-contained. Discoveries in one area find applications in others, often unpredictably. Various notable examples of this involve astronomical studies. Isaac Newton's laws of motion and gravity emerged from the analysis of planetary and lunar orbits. Observations during the 1919 solar eclipse provided dramatic confirmation of Albert Einstein's general theory of relativity, which gained further support with the discovery of the binary pulsar designated PSR 1913+16 and the observation of gravity waves from merging black holes and neutron stars. The behaviour of nuclear matter and of some elementary particles is now better understood as a result of measurements of neutron stars and the cosmological helium abundance, respectively. Study of the theory of synchrotron radiation was greatly stimulated by the detection of polarized visible radiation emitted by high-energy electrons in the supernova remnant known as the Crab Nebula. Dedicated particle accelerators are now being used to produce synchrotron radiation to probe the structure of solid materials and make detailed X-ray images of tiny samples, including biological structures.

Astronomical knowledge also has had a broad impact beyond science. The earliest calendars were based on astronomical observations of the cycles of repeated solar and lunar positions. Also, for centuries, familiarity with the positions and apparent motions of the stars through the seasons enabled sea voyagers to navigate with moderate accuracy. Perhaps the single greatest effect that astronomical studies have had on our modern society has been in molding its perceptions and opinions. Our conceptions of the cosmos and our place in it, our perceptions of space and time, and the development of the systematic pursuit of knowledge known as the scientific method have been profoundly influenced by astronomical observations. In addition, the power of science to provide the basis for accurate predictions of such phenomena as eclipses and the positions of the planets and later, so dramatically, of comets has shaped an attitude toward science that remains an important social force today.

Importance of Earth Science

The significance of this is understood by knowing the regions that are covered by the different branches of it. Since the study of the oceans that covers about seventy-one percent of the surfaces of our planet. It serves one of the most important divisions of the science. The Earth is the only planet which supports life and the only planet where life is said to be continuously sustained. Oceans are considered as the areas of the origin of life on the earth and a major determinant of the earth's atmospheric condition that serves as determinants of various life processes on various parts of the earth. Also, the study of the different fossil forms that are present under the earth's surface gives us the information about the forms of life present at a geological sense and it also known to establish a bond between the ancestral & the living forms. The Rocks that are found in the variety of parts of the earth provides data about the evolution of rocks at ancient times. Hence, it is one among the principal branches of physics to study.

References

- Earth, place: britannica.com, Retrieved 2 March, 2019

- What-is-earth-science: geology.com, Retrieved 14 June, 2019

- What-is-geology: geologypage.com, Retrieved 5 August, 2019

- Important-principles-of-geology: newworldencyclopedia.org, Retrieved 15 February, 2019

- Outcome-what-is-geology: lumenlearning.com, Retrieved 25 January, 2019

- Geography: worldatlas.com, Retrieved 3 April, 2019

- Physical-oceanography: researchgate.net, Retrieved 28 February, 2019

- Science-environmental, environment: brighthub.com, Retrieved 8 July, 2019

- Astronomy, science: britannica.com, Retrieved 11 May, 2019

2

Earth: Structure and Properties

The internal structure of the Earth is made up of several layers, namely, the crust, the mantle and the core. Crust is further sub-divided into oceanic crust and continental crust. The topics elaborated in this chapter will help in gaining a better perspective about these components of the Earth's structure as well as their properties.

Structure of The Earth

The structure of the Earth is divided into 3 layers.

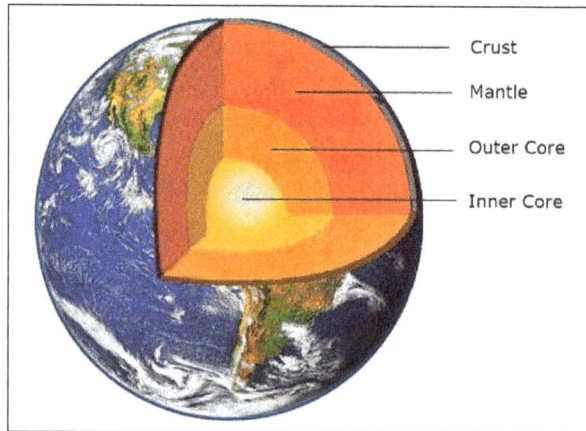

The Crust

The Earth's crust is an extremely thin layer of rock that makes up the outermost solid shell of our planet. In relative terms, it's thickness is like that of the skin of an apple. It amounts to less than half of 1 percent of the planet's total mass but plays a vital role in most of Earth's natural cycles.

The crust can be thicker than 80 kilometers in some spots and less than one kilometer thick in others. Underneath it lies the mantle, a layer of silicate rock approximately 2700 kilometers thick. The mantle accounts for the bulk of the Earth.

The crust is composed of many different types of rocks that fall into three main categories: igneous, metamorphic and sedimentary. However, most of those rocks originated as either granite or basalt. The mantle beneath is made of peridotite. Bridgmanite, the most common mineral on Earth, is found in the deep mantle.

Oceanic Crust

Oceanic crust is the outermost layer of Earth's lithosphere that is found under the oceans and formed at spreading centres on oceanic ridges, which occur at divergent plate boundaries.

Oceanic crust is about 6 km (4 miles) thick. It is composed of several layers, not including the overlying sediment. The topmost layer, about 500 metres (1,650 feet) thick, includes lavas made of basalt (that is, rock material consisting largely of plagioclase [feldspar] and pyroxene). Oceanic crust differs from continental crust in several ways: it is thinner, denser, younger, and of different chemical composition. Like continental crust, however, oceanic crust is destroyed in subduction zones.

The lavas are generally of two types: pillow lavas and sheet flows. Pillow lavas appear to be shaped exactly as the name implies—like large overstuffed pillows about 1 metre (3 feet) in cross section and 1 to several metres long. They commonly form small hills tens of metres high at the spreading centres. Sheet flows have the appearance of wrinkled bed sheets. They commonly are thin (only about 10 cm [4 inches] thick) and cover a broader area than pillow lavas. There is evidence that sheet flows are erupted at higher temperatures than those of the pillow variety. On the East Pacific Rise at 8° S latitude, a series of sheet flow eruptions (possibly since the mid-1960s) have covered more than 220 square km (85 square miles) of seafloor to an average depth of 70 metres (230 feet).

Below the lava is a layer composed of feeder, or sheeted, dikes that measures more than 1 km (0.6 mile) thick. Dikes are fractures that serve as the plumbing system for transporting magmas (molten rock material) to the seafloor to produce lavas. They are about 1 metre (3 feet) wide, subvertical, and elongate along the trend of the spreading centre where they formed, and they abut one another's sides—hence the term sheeted. These dikes also are of basaltic composition. There are two layers below the dikes totaling about 4.5 km (3 miles) in thickness. Both of these include gabbros, which are essentially basalts with coarser mineral grains. These gabbro layers are thought to represent the magma chambers, or pockets of lava, that ultimately erupt on the seafloor. The upper gabbro layer is isotropic (uniform) in structure. In some places this layer includes pods of plagiogranite, a differentiated rock richer in silica than gabbro. The lower gabbro layer has a stratified structure and evidently represents the floor or sides of the magma chamber. This layered structure is called cumulate, meaning that the layers (which measure up to several metres thick) result from the sedimentation of minerals out of the liquid magma. The layers in the cumulate gabbro have less silica but are richer in iron and magnesium than the upper portions of the crust. Olivine, an iron-magnesium silicate, is a common mineral in the lower gabbro layer.

Crew members aboard a drilling ship inspecting a rock core during a scientific
expedition that succeeded for the first time in drilling through the upper oceanic crust.

The oceanic crust lies atop Earth's mantle, as does the continental crust. Mantle rock
is composed mostly of peridotite, which consists primarily of the mineral olivine with
small amounts of pyroxene and amphibole.

Life Cycle of the Oceanic Crust

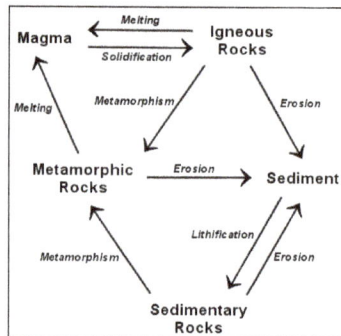

Figure: Diagram of the rock cycle.

All rocks in Earth's crust are constantly being recycled through the rock cycle. The rock
cycle is the transition of rocks among three different rock types over millions of years
of geologic time. Igneous rock is formed by the cooling and crystallization of molten
magma at volcanoes and mid-ocean ridges, where new crust is generated. Examples
of igneous rock are basalt, granite, and andesite. Over time, igneous rocks may experience weathering and erosion from exposure to water and the atmosphere to produce
sediments. The deposition and hardening of these sediments forms sedimentary rocks.
Both igneous and sedimentary rock types can transform physically and chemically into
a third rock type. Metamorphic rocks are formed when igneous or sedimentary rocks
are exposed to conditions of high heat and pressure. Examples of metamorphic rock
include marble, slate, schist, and gneiss. Metamorphic rocks can also transform to sedimentary rocks through weathering, erosion, and sediment deposition.

(A) Basalt, an example of igneous rock, from Mauna Ulu Lava Field, East Rift Zone, Kilauea Volcano, Hawaii.

(B) Sandstone, an example of sedimentary rock, Jackson County, Ohio.

(C) Marble, an example of metamorphic rock, Czech Republic.

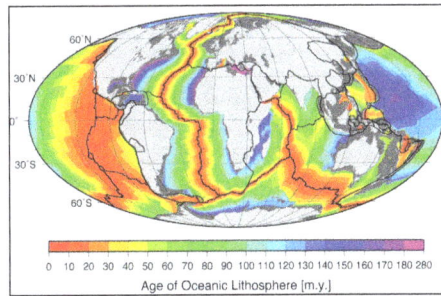
The age of oceanic crust in millions of years. The youngest crust (shown in red) is near mid ocean ridges and spreading zones.

All three rock types in the earth's crust—igneous, sedimentary, and metamorphic—can also be recycled back to their original molten magma form. This process occurs when oceanic crust is pushed back into the mantle at subduction zones. As old oceanic crust is subducted and melted into magma, new oceanic crust in the form of igneous rock is formed at mid-ocean ridges and volcanic hotspots. This recycling accounts for the recycling of 60 percent of Earth's surface every 200 million years, making the oldest recorded oceanic crust rock roughly the same age. Because of this recycling, the age of the oceanic crust varies depending on location. Areas where new crust is being formed at mid-ocean ridges are much younger than zones further away. By contrast, continental crust is rarely recycled and is typically much older. The oldest recorded rocks on Earth are all located on continental crust in northern Canada and western Australia and date to approximately 3.8 to 4.4 billion years old.

Continental Crust

Continental crust is the solid, outermost layer of the Earth, lying above the mantle. This layer is sometimes called sial because its bulk composition is more felsic compared to the oceanic crust, called sima which has a more mafic bulk composition. The crust that includes continents is called continental crust and is about 35.4 to 70 km (22 to 43.4 mi) thick. It consists mostly of rocks, such as granites and granodiorites that are rich in silica and aluminum, with minor amounts of iron, magnesium, calcium, sodium, and potassium.

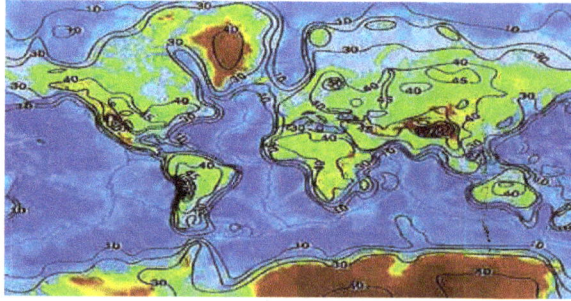

Consisting mostly of granitic rock, it has a density of about 2.7g/cm3 and is less dense than the material of the Earth's mantle, which consists of mafic rock. Continental crust is also less dense than oceanic crust, though it is considerably thicker; mostly 35 to 40 km versus the average oceanic thickness of around 7-10 km. About 40% of the Earth's surface is now underlain by continental crust.

The crust that includes ocean floors is called oceanic crust and is about 4.8 to 9.7 km (3 to 6 mi) thick. It has a similar composition to that of continental crust, but has higher concentrations of iron, magnesium, and calcium and is denser than continental crust. The predominant type of rock in oceanic crust is basalt.

Formation and Forces at Work of Continental Crust

There are two types of crust: continental and oceanic. Continental crust is the crust under which the continents are built and is 10-70 km thick, while oceanic crust is the crust under the oceans, and is only 5-7 km thick. The crust isn't one large piece; rather, it's more like an egg shell that has been cracked repeatedly. The crust contains sections, called plates. Just like the floatation devices resting on top of a pool of water, these plates are continually on the move. When they hit each other, slide along each other, or move away from each other, you can get earthquakes and volcanoes.

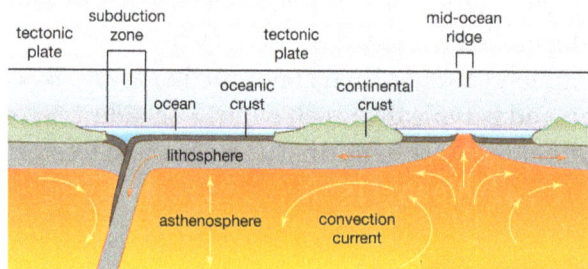

Continental crust is more complex than oceanic crust in its structure and origin and is formed primarily at subduction zones. Lateral growth occurs by the addition of rock scraped off the top of oceanic plates as they are subducted beneath continental margins. The continental crust formed through these interactions between plates. When two plates collide, one plate can subduct beneath the other. Subduction is simply the process by which one plate moves underneath another plate.

Continental crust resists subduction. Consequently, the mean age of the continents is almost two billion years, more than 30 times the average age of the oceanic crust. Thus, continents are the prime repositories of information concerning Earth's geologic evolution, but understanding their formation requires knowledge of processes in the ocean basins from which they evolved.

Today there are ~7 billion cubic kilometers of continental crust, but in the past there may have been fewer or more. The relative permanence of continental crust contrasts with the short life of oceanic crust. As a consequence of the density difference, when active margins of continental crust meet oceanic crust in subduction zones, the oceanic crust is typically subducted back into the mantle. Because of its relative low density, continental crust is only rarely subducted or re-cycled back into the mantle.

The height of mountain ranges is usually related to the thickness of crust. This result from the isostasy associated with orogeny (mountain formation). The crust is thickened by the compressive forces related to subduction or continental collision. The high temperatures and pressures at depth, often combined with a long history of complex distortion, cause much of the lower continental crust to be metamorphic – the main exception to this being recent igneous intrusions. Igneous rock may also be "underplated" to the underside of the crust, i.e. adding to the crust by forming a layer immediately beneath it.

Importance of Continental Crust

There is little evidence of continental crust prior to 3.5 Ga, and there was relatively rapid development on shield areas consisting of continental crust between 3.0 and 2.5 Ga. All continental crust ultimately derives from the fractional differentiation of oceanic crust over many eons. This process has been and continues today primarily as a result of the volcanism associated with subduction. Because the surface of continental crust mainly lies above sea level, its existence allowed land life to evolve from marine life. Its existence also provides broad expanses of shallow water known as epeiric seas and continental shelves where complex metazoan life could become established during early Paleozoic time, in what is now called the Cambrian explosion.

Today there are ~7 billion cubic kilometers of continental crust, but in the past there may have been less or more. Today continental crust is produced and destroyed mostly by plate tectonic processes, especially at convergent plate boundaries. New material can be added to the continents by the partial melting of oceanic crust at subduction zones, causing the lighter material to rise as magma, forming volcanoes. Also, material can be accreted "horizontally" when volcanic island arcs, seamounts or similar structures collide with the side of the continent as a result of plate tectonic movements. Continental crust is also lost, due to erosion and sediment subduction, tectonic erosion of forearcs, delamination, and deep subduction of continental crust in collision zones.

The Mantle

The mantle is the mostly-solid bulk of Earth's interior. The mantle lies between Earth's dense, super-heated core and its thin outer layer, the crust. The mantle is about 2,900 kilometers (1,802 miles) thick, and makes up a whopping 84% of Earth's total volume.

As Earth began to take shape about 4.5 billion years ago, iron and nickel quickly separated from other rocks and minerals to form the core of the new planet. The molten material that surrounded the core was the early mantle.

Over millions of years, the mantle cooled. Water trapped inside minerals erupted with lava, a process called "outgassing." As more water was outgassed, the mantle solidified.

The rocks that make up Earth's mantle are mostly silicates—a wide variety of compounds that share a silicon and oxygen structure. Common silicates found in the mantle include olivine, garnet, and pyroxene. The other major type of rock found in the mantle is magnesium oxide. Other mantle elements include iron, aluminum, calcium, sodium, and potassium.

The temperature of the mantle varies greatly, from 1000 °Celsius (1832 °Fahrenheit) near its boundary with the crust, to 3700 °Celsius (6692 °Fahrenheit) near its boundary with the core. In the mantle, heat and pressure generally increase with depth. The geothermal gradient is a measurement of this increase. In most places, the geothermal gradient is about 25 °Celsius per kilometer of depth (1 °Fahrenheit per 70 feet of depth).

The viscosity of the mantle also varies greatly. It is mostly solid rock, but less viscous at tectonic plate boundaries and mantle plumes. Mantle rocks there are soft and able to move plastically (over the course of millions of years) at great depth and pressure.

The transfer of heat and material in the mantle helps determine the landscape of Earth. Activity in the mantle drives plate tectonics, contributing to volcanoes, seafloor spreading, earthquakes, and orogeny (mountain-building).

The mantle is divided into several layers: the upper mantle, the transition zone, the lower mantle, and D" (D double-prime), the strange region where the mantle meets the outer core.

Upper Mantle

The upper mantle extends from the crust to a depth of about 410 kilometers (255 miles). The upper mantle is mostly solid, but its more malleable regions contribute to tectonic activity.

Two parts of the upper mantle are often recognized as distinct regions in Earth's interior: the lithosphere and the asthenosphere.

Lithosphere

The lithosphere is the solid, outer part of the Earth, extending to a depth of about 100 kilometers (62 miles). The lithosphere includes both the crust and the brittle upper portion of the mantle. The lithosphere is both the coolest and the most rigid of Earth's layers.

The most well-known feature associated with Earth's lithosphere is tectonic activity. Tectonic activity describes the interaction of the huge slabs of lithosphere called tectonic plates. The lithosphere is divided into 15 major tectonic plates: the North American, Caribbean, South American, Scotia, Antarctic, Eurasian, Arabian, African, Indian, Philippine, Australian, Pacific, Juan de Fuca, Cocos, and Nazca.

The division in the lithosphere between the crust and the mantle is called the Mohorovicic discontinuity, or simply the Moho. The Moho does not exist at a uniform depth, because not all regions of Earth are equally balanced in isostatic equilibrium. Isostasy describes the physical, chemical, and mechanical differences that allow the crust to "float" on the sometimes more malleable mantle. The Moho is found at about 8 kilometers (5 miles) beneath the ocean and about 32 kilometers (20 miles) beneath continents.

Different types of rocks distinguish lithospheric crust and mantle. Lithospheric crust is characterized by gneiss (continental crust) and gabbro (oceanic crust). Below the Moho, the mantle is characterized by peridotite, a rock mostly made up of the minerals olivine and pyroxene.

Asthenosphere

The asthenosphere is the denser, weaker layer beneath the lithospheric mantle. It lies between about 100 kilometers (62 miles) and 410 kilometers (255 miles) beneath Earth's surface. The temperature and pressure of the asthenosphere are so high that rocks soften and partly melt, becoming semi-molten.

The asthenosphere is much more ductile than either the lithosphere or lower mantle. Ductility measures a solid material's ability to deform or stretch under stress. The asthenosphere is generally more viscous than the lithosphere, and the lithosphere-asthenosphere boundary (LAB) is the point where geologists and rheologists—scientists who study the flow of matter—mark the difference in ductility between the two layers of the upper mantle.

The very slow motion of lithospheric plates "floating" on the asthenosphere is the cause of plate tectonics, a process associated with continental drift, earthquakes, the formation of mountains, and volcanoes. In fact, the lava that erupts from volcanic fissures is actually the asthenosphere itself, melted into magma.

Of course, tectonic plates are not really floating, because the asthenosphere is not liquid. Tectonic plates are only unstable at their boundaries and hot spots.

Transition Zone

From about 410 kilometers (255 miles) to 660 kilometers (410 miles) beneath Earth's surface, rocks undergo radical transformations. This is the mantle's transition zone.

In the transition zone, rocks do not melt or disintegrate. Instead, their crystalline structure changes in important ways. Rocks become much, much more dense.

The transition zone prevents large exchanges of material between the upper and lower mantle. Some geologists think that the increased density of rocks in the transition zone prevents subducted slabs from the lithosphere from falling further into the mantle. These huge pieces of tectonic plates stall in the transition zone for millions of years before mixing with other mantle rock and eventually returning to the upper mantle as part of the asthenosphere, erupting as lava, becoming part of the lithosphere, or emerging as new oceanic crust at sites of seafloor spreading.

Some geologists and rheologists, however, think subducted slabs can slip beneath the transition zone to the lower mantle. Other evidence suggests that the transition layer is permeable, and the upper and lower mantle exchange some amount of material.

Water

Perhaps the most important aspect of the mantle's transition zone is its abundance of water. Crystals in the transition zone hold as much water as all the oceans on Earth's surface.

Water in the transition zone is not "water" as we know it. It is not liquid, vapor, solid, or even plasma. Instead, water exists as hydroxide. Hydroxide is an ion of hydrogen and oxygen with a negative charge. In the transition zone, hydroxide ions are trapped in the crystalline structure of rocks such as ringwoodite and wadsleyite. These minerals are formed from olivine at very high temperatures and pressure.

Near the bottom of the transition zone, increasing temperature and pressure transform ringwoodite and wadsleyite. Their crystal structures are broken and hydroxide escapes as "melt." Melt particles flow upwards, toward minerals that can hold water. This allows the transition zone to maintain a consistent reservoir of water.

Geologists and rheologists think that water entered the mantle from Earth's surface during subduction. Subduction is the process in which a dense tectonic plate slips or melts beneath a more buoyant one. Most subduction happens as an oceanic plate slips beneath a less-dense plate. Along with the rocks and minerals of the lithosphere, tons of water and carbon are also transported to the mantle. Hydroxide and water are returned to the upper mantle, crust, and even atmosphere through mantle convection, volcanic eruptions, and seafloor spreading.

Lower Mantle

The lower mantle extends from about 660 kilometers (410 miles) to about 2,700 kilometers (1,678 miles) beneath Earth's surface. The lower mantle is hotter and denser than the upper mantle and transition zone. The lower mantle is much less ductile than the upper mantle and transition zone. Although heat usually corresponds to softening rocks, intense pressure keeps the lower mantle solid.

Geologists do not agree about the structure of the lower mantle. Some geologists think that subducted slabs of lithosphere have settled there. Other geologists think that the lower mantle is entirely unmoving and does not even transfer heat by convection.

D Double-Prime

Beneath the lower mantle is a shallow region called D", or "d double-prime." In some areas, D" is a nearly razor-thin boundary with the outer core. In other areas, D" has thick accumulations of iron and silicates. In still other areas, geologists and seismologists have detected areas of huge melt.

The unpredictable movement of materials in D" is influenced by the lower mantle and outer core. The iron of the outer core influences the formation of a diapir, a dome-shaped geologic feature (igneous intrusion) where more fluid material is forced into brittle overlying rock. The iron diapir emits heat and may release a huge, bulging pulse of either material or energy—just like a Lava Lamp. This energy blooms upward, transferring heat to the lower mantle and transition zone, and maybe even erupting as a mantle plume. At the base of the mantle, about 2,900 kilometers (1,802 miles) below the surface, is the core-mantle boundary, or CMB. This point, called the Gutenberg discontinuity, marks the end of the mantle and the beginning of Earth's liquid outer core.

Mantle Convection

Mantle convection describes the movement of the mantle as it transfers heat from the white-hot core to the brittle lithosphere. The mantle is heated from below, cooled from above, and its overall temperature decreases over long periods of time. All these elements contribute to mantle convection.

Convection currents transfer hot, buoyant magma to the lithosphere at plate boundaries and hot spots. Convection currents also transfer denser, cooler material from the crust to Earth's interior through the process of subduction. Earth's heat budget, which measures the flow of thermal energy from the core to the atmosphere, is dominated by mantle convection. Earth's heat budget drives most geologic processes on Earth, although its energy output is dwarfed by solar radiation at the surface.

Geologists debate whether mantle convection is "whole" or "layered." Whole-mantle convection describes a long, long recycling process involving the upper mantle, transition zone, lower mantle, and even D". In this model, the mantle convects in a single process. A subducted slab of lithosphere may slowly slip into the upper mantle and fall to the transition zone due to its relative density and coolness. Over millions of years, it may sink further into the lower mantle. Convection currents may then transport the hot, buoyant material in D" back through the other layers of the mantle. Some of that material may even emerge as lithosphere again, as it is spilled onto the crust through volcanic eruptions or seafloor spreading.

Layered-mantle convection describes two processes. Plumes of superheated mantle material may bubble up from the lower mantle and heat a region in the transition zone before falling back. Above the transition zone, convection may be influenced by heat transferred from the lower mantle as well as discrete convection currents in the upper mantle driven by subduction and seafloor spreading. Mantle plumes emanating from the upper mantle may gush up through the lithosphere as hot spots.

Mantle Plumes

A mantle plume is an upwelling of superheated rock from the mantle. Mantle plumes are the likely cause of "hot spots," volcanic regions not created by plate tectonics. As a mantle plume reaches the upper mantle, it melts into a diapir. This molten material heats the asthenosphere and lithosphere, triggering volcanic eruptions. These volcanic eruptions make a minor contribution to heat loss from Earth's interior, although tectonic activity at plate boundaries is the leading cause of such heat loss.

The Hawaiian hot spot, in the middle of the North Pacific Ocean, sits above a likely mantle plume. As the Pacific plate moves in a generally northwestern motion, the Hawaiian hot spot remains relatively fixed. Geologists think this has allowed the Hawaiian hot spot to create a series of volcanoes, from the 85-million-year-old Meiji Seamount near Russia's Kamchatka Peninsula, to the Loihi Seamount, a submarine volcano southeast of the "Big Island" of Hawaii. Loihi, a mere 400,000 years old, will eventually become the newest Hawaiian island.

Geologists have identified two so-called "superplumes." These superplumes, or large low shear velocity provinces (LLSVPs), have their origins in the melt material of D". The Pacific LLSVP influences geology throughout most of the southern Pacific Ocean (including the Hawaiian hot spot). The African LLSVP influences the geology through-out most of southern and western Africa. Geologists think mantle plumes may be in-fluenced by many different factors. Some may pulse, while others may be heated con-tinually. Some may have a single diapir, while others may have multiple "stems." Some mantle plumes may arise in the middle of a tectonic plate, while others may be "cap-tured" by seafloor spreading zones.

Some geologists have identified more than a thousand mantle plumes. Some geologists think mantle plumes don't exist at all. Until tools and technology allow geologists to more thoroughly explore the mantle, the debate will continue.

Exploring the Mantle

The mantle has never been directly explored. Even the most sophisticated drilling equipment has not reached beyond the crust. Drilling all the way down to the Moho (the division between the Earth's crust and mantle) is an important scientific mile-stone, but despite decades of effort, nobody has yet succeeded. In 2005, scientists with the Integrated Ocean Drilling Project drilled 1,416 meters (4,644 feet) below the North Atlantic seafloor and claimed to have come within just 305 meters (1,000 feet) of the Moho.

Xenoliths

Many geologists study the mantle by analyzing xenoliths. Xenoliths are a type of intru-sion—a rock trapped inside another rock. The xenoliths that provide the most informa-tion about the mantle are diamonds. Diamonds form under very unique conditions: in the upper mantle, at least 150 kilometers (93 miles) beneath the surface. Above depth and pressure, the carbon crystallizes as graphite, not diamond. Diamonds are brought to the surface in explosive volcanic eruptions, forming "diamond pipes" of rocks called kimberlites and lamprolites.

The diamonds themselves are of less interest to geologists than the xenoliths some con-tain. These intrusions are minerals from the mantle, trapped inside the rock-hard dia-mond. Diamond intrusions have allowed scientists to glimpse as far as 700 kilometers (435 miles) beneath Earth's surface—the lower mantle. Xenolith studies have revealed that rocks in the deep mantle are most likely 3-billion-year old slabs of subducted sea-floor. The diamond intrusions include water, ocean sediments, and even carbon.

Seismic Waves

Most mantle studies are conducted by measuring the spread of shock waves from earth-quakes, called seismic waves. The seismic waves measured in mantle studies are called body waves, because these waves travel through the body of the Earth. The velocity of body waves differs with density, temperature, and type of rock.

There are two types of body waves: primary waves, or P-waves, and secondary waves, or S-waves. P-waves, also called pressure waves, are formed by compressions. Sound waves are P-waves—seismic P-waves are just far too low a frequency for people to hear. S-waves, also called shear waves, measure motion perpendicular to the energy transfer. S-waves are unable to transmit through fluids or gases.

Instruments placed around the world measure these waves as they arrive at different points on the Earth's surface after an earthquake. P-waves (primary waves) usually arrive first, while s-waves arrive soon after. Both body waves "reflect" off different types of rocks in different ways. This allows seismologists to identify different rocks present in Earth's crust and mantle far beneath the surface. Seismic reflections, for instance, are used to identify hidden oil deposits deep below the surface. Sudden, predictable changes in the velocities of body waves are called "seismic discontinuities." The Moho is a discontinuity marking the boundary of the crust and upper mantle. The so-called "410-kilometer discontinuity" marks the boundary of the transition zone.

The Gutenberg discontinuity is more popularly known as the core-mantle boundary (CMB). At the CMB, S-waves, which can't continue in liquid, suddenly disappear, and P-waves are strongly refracted, or bent. This alerts seismologists that the solid and molten structure of the mantle has given way to the fiery liquid of the outer core.

Mantle Maps

Cutting-edge technology has allowed modern geologists and seismologists to produce mantle maps. Most mantle maps display seismic velocities, revealing patterns deep below Earth's surface. Geoscientists hope that sophisticated mantle maps can plot the body waves of as many as 6,000 earthquakes with magnitudes of at least 5.5. These mantle maps may be able to identify ancient slabs of subducted material and the precise position and movement of tectonic plates. Many geologists think mantle maps may even provide evidence for mantle plumes and their structure.

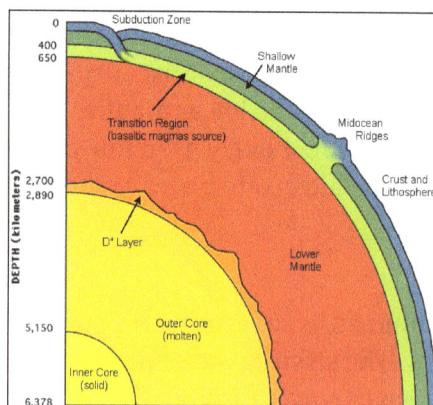

The mantle, between the brittle crust and super-dense core, makes up a whopping 84% of Earth's total volume.

The Core

Earth's core is the very hot, very dense center of our planet. The ball-shaped core lies beneath the cool, brittle crust and the mostly-solid mantle. The core is found about 2,900 kilometers (1,802 miles) below Earth's surface, and has a radius of about 3,485 kilometers (2,165 miles).

Planet Earth is older than the core. When Earth was formed about 4.5 billion years ago, it was a uniform ball of hot rock. Radioactive decay and leftover heat from planetary formation (the collision, accretion, and compression of space rocks) caused the ball to get even hotter. Eventually, after about 500 million years, our young planet's temperature heated to the melting point of iron—about 1,538 °Celsius (2,800 °Fahrenheit). This pivotal moment in Earth's history is called the iron catastrophe.

The iron catastrophe allowed greater, more rapid movement of Earth's molten, rocky material. Relatively buoyant material, such as silicates, water, and even air, stayed close to the planet's exterior. These materials became the early mantle and crust. Droplets of iron, nickel, and other heavy metals gravitated to the center of Earth, becoming the early core. This important process is called planetary differentiation.

Earth's core is the furnace of the geothermal gradient. The geothermal gradient measures the increase of heat and pressure in Earth's interior. The geothermal gradient is about 25 °Celsius per kilometer of depth (1 °Fahrenheit per 70 feet). The primary contributors to heat in the core are the decay of radioactive elements, leftover heat from planetary formation, and heat released as the liquid outer core solidifies near its boundary with the inner core.

Unlike the mineral-rich crust and mantle, the core is made almost entirely of metal—specifically, irons and nickel. The shorthand used for the core's iron-nickel alloys is simply the elements' chemical symbols—NiFe. Elements that dissolve in iron, called siderophiles, are also found in the core. Because these elements are found much more rarely on Earth's crust, many siderophiles are classified as "precious metals." Siderophile elements include gold, platinum, and cobalt.

Another key element in Earth's core is sulfur—in fact 90% of the sulfur on Earth is found in the core. The confirmed discovery of such vast amounts of sulfur helped explain a geologic mystery: If the core was primarily NiFe, why wasn't it heavier? Geoscientists speculated that lighter elements such as oxygen or silicon might have been present. The abundance of sulfur, another relatively light element, explained the conundrum.

Although we know that the core is the hottest part of our planet, its precise temperatures are difficult to determine. The fluctuating temperatures in the core depend on pressure, the rotation of the Earth, and the varying composition of core elements. In general, temperatures range from about 4,400 °Celsius (7,952 °Fahrenheit) to about 6,000 °Celsius (10,800 °Fahrenheit).

The core is made of two layers: the outer core, which borders the mantle, and the inner core. The boundary separating these regions is called the Bullen discontinuity.

Outer Core

The outer core, about 2,200 kilometers (1,367 miles) thick, is mostly composed of liquid iron and nickel. The NiFe alloy of the outer core is very hot, between 4,500° and 5,500 °Celsius (8,132° and 9,932 °Fahrenheit). The liquid metal of the outer core has very low viscosity, meaning it is easily deformed and malleable. It is the site of violent convection. The churning metal of the outer core creates and sustains Earth's magnetic field. The hottest part of the core is actually the Bullen discontinuity, where temperatures reach 6,000 °Celsius (10,800 °Fahrenheit)—as hot as the surface of the sun.

Inner Core

The inner core is a hot, dense ball of (mostly) iron. It has a radius of about 1,220 kilometers (758 miles). Temperature in the inner core is about 5,200 °Celsius (9,392 °Fahrenheit). The pressure is nearly 3.6 million atmosphere (atm).

The temperature of the inner core is far above the melting point of iron. However, unlike the outer core, the inner core is not liquid or even molten. The inner core's intense pressure—the entire rest of the planet and its atmosphere—prevents the iron from melting. The pressure and density are simply too great for the iron atoms to move into a liquid state. Because of this unusual set of circumstances, some geophysicists prefer to interpret the inner core not as a solid, but as a plasma behaving as a solid.

The liquid outer core separates the inner core from the rest of the Earth, and as a result, the inner core rotates a little differently than the rest of the planet. It rotates eastward, like the surface, but it's a little faster, making an extra rotation about every 1,000 years.

Geoscientists think that the iron crystals in the inner core are arranged in an "hcp" (hexagonal close-packed) pattern. The crystals align north-south, along with Earth's axis of rotation and magnetic field.

The orientation of the crystal structure means that seismic waves—the most reliable way to study the core—travel faster when going north-south than when going east-west. Seismic waves travel four seconds faster pole-to-pole than through the Equator.

Growth in the Inner Core

As the entire Earth slowly cools, the inner core grows by about a millimeter every year. The inner core grows as bits of the liquid outer core solidify or crystallize. Another word for this is "freezing," although it's important to remember that iron's freezing point more than 1,000 °Celsius (1,832 °Fahrenheit). The growth of the inner core is not uniform. It occurs in lumps and bunches, and is influenced by activity in the mantle.

Growth is more concentrated around subduction zones—regions where tectonic plates are slipping from the lithosphere into the mantle, thousands of kilometers above the core. Subducted plates draw heat from the core and cool the surrounding area, causing increased instances of solidification.

Growth is less concentrated around "superplumes" or LLSVPs. These ballooning masses of superheated mantle rock likely influence "hot spot" volcanism in the lithosphere, and contribute to a more liquid outer core.

The core will never "freeze over." The crystallization process is very slow, and the constant radioactive decay of Earth's interior slows it even further. Scientists estimate it would take about 91 billion years for the core to completely solidify—but the sun will burn out in a fraction of that time (about 5 billion years).

Core Hemispheres

Just like the lithosphere, the inner core is divided into eastern and western hemispheres. These hemispheres don't melt evenly, and have distinct crystalline structures. The western hemisphere seems to be crystallizing more quickly than the eastern hemisphere. In fact, the eastern hemisphere of the inner core may actually be melting.

Geoscientists recently discovered that the inner core itself has a core—the inner core. This strange feature differs from the inner core in much the same way the inner core differs from the outer core. Scientists think that a radical geologic change about 500 million years ago caused this inner inner core to develop.

The crystals of the inner core are oriented east-west instead of north-south. This orientation is not aligned with either Earth's rotational axis or magnetic field. Scientists think the iron crystals may even have a completely different structure (not hcp), or exist at a different phase.

Magnetism

Earth's magnetic field is created in the swirling outer core. Magnetism in the outer core is about 50 times stronger than it is on the surface.

It might be easy to think that Earth's magnetism is caused by the big ball of solid iron in the middle. But in the inner core, the temperature is so high the magnetism of iron is altered. Once this temperature, called the Curie point, is reached, the atoms of a substance can no longer align to a magnetic point.

Dynamo Theory

Some geoscientists describe the outer core as Earth's "geodynamo." For a planet to have a geodynamo, it must rotate, it must have a fluid medium in its interior, the fluid must be able to conduct electricity, and it must have an internal energy supply that drives convection in the liquid.

Variations in rotation, conductivity, and heat impact the magnetic field of a geodynamo. Mars, for instance, has a totally solid core and a weak magnetic field. Venus has a liquid core, but rotates too slowly to churn significant convection currents. It, too, has a weak magnetic field. Jupiter, on the other hand, has a liquid core that is constantly swirling due to the planet's rapid rotation.

Earth is the "Goldilocks" geodynamo. It rotates steadily, at a brisk 1,675 kilometers per hour (1,040 miles per hour) at the Equator. Coriolis forces, an artifact of Earth's rotation, cause convection currents to be spiral. The liquid iron in the outer core is an excellent electrical conductor, and creates the electrical currents that drive the magnetic field.

The energy supply that drives convection in the outer core is provided as droplets of liquid iron freeze onto the solid inner core. Solidification releases heat energy. This heat, in turn, makes the remaining liquid iron more buoyant. Warmer liquids spiral upward, while cooler solids spiral downward under intense pressure convection.

Earth's Magnetic Field

Earth's magnetic field is crucial to life on our planet. It protects the planet from the charged particles of the solar wind. Without the shield of the magnetic field, the solar wind would strip Earth's atmosphere of the ozone layer that protects life from harmful ultraviolet radiation.

Although Earth's magnetic field is generally stable, it fluctuates constantly. As the liquid outer core moves, for instance, it can change the location of the magnetic North and South Poles. The magnetic North Pole moves up to 64 kilometers (40 miles) every year.

Fluctuations in the core can cause Earth's magnetic field to change even more dramatically. Geomagnetic pole reversals, for instance, happen about every 200,000 to 300,000 years. Geomagnetic pole reversals are just what they sound like: a change in the planet's magnetic poles, so that the magnetic North and South Poles are reversed. These "pole flips" are not catastrophic—scientists have noted no real changes in plant or animal life, glacial activity, or volcanic eruptions during previous geomagnetic pole reversals.

Studying the Core

Geoscientists cannot study the core directly. All information about the core has come from sophisticated reading of seismic data, analysis of meteorites, lab experiments with temperature and pressure, and computer modeling.

Most core research has been conducted by measuring seismic waves, the shock waves released by earthquakes at or near the surface. The velocity and frequency of seismic body waves changes with pressure, temperature, and rock composition.

In fact, seismic waves helped geoscientists identify the structure of the core itself. In the late 19th century, scientists noted a "shadow zone" deep in the Earth, where a type of body wave called an s-wave either stopped entirely or was altered. S-waves are unable to transmit through fluids or gases. The sudden "shadow" where s-waves disappeared indicated that Earth had a liquid layer.

In the 20th century, geoscientists discovered an increase in the velocity of p-waves, another type of body wave, at about 5,150 kilometers (3,200 miles) below the surface. The increase in velocity corresponded to a change from a liquid or molten medium to a solid. This proved the existence of a solid inner core.

Meteorites, space rocks that crash to Earth, also provide clues about Earth's core. Most meteorites are fragments of asteroids, rocky bodies that orbit the sun between Mars and Jupiter. Asteroids formed about the same time, and from about the same material, as Earth. By studying iron-rich chondrite meteorites, geoscientists can get a peek into the early formation of our solar system and Earth's early core.

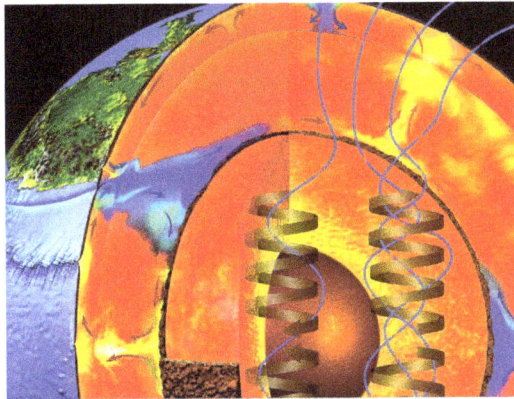

The core is the hottest, densest part of the Earth.

In the lab, the most valuable tool for studying forces and reactions at the core is the diamond anvil cell. Diamond anvil cells use the hardest substance on Earth (diamonds) to simulate the incredibly high pressure at the core. The device uses an x-ray laser to simulate the core's temperature. The laser is beamed through two diamonds squeezing a sample between them.

Complex computer modeling has also allowed scientists to study the core. In the 1990s, for instance, modeling beautifully illustrated the geodynamo—complete with pole flips.

References

- Composition-of-the-earth: science4fun.info, Retrieved 15 January, 2019

- All-about-the-earths-crust: thoughtco.com, Retrieved 5 March, 2019

- Oceanic-crust: britannica.com, Retrieved 3 May, 2019

- Node, exploringourfluidearth: hawaii.edu, Retrieved 23 February, 2019

- Continental-crust, geography, science: assignmentpoint.com, Retrieved 19 April, 2019

3

Earth's Spheres

Earth's spheres refer to the subsystems which make up the natural environment of the Earth. The four major spheres of the Earth are atmosphere, biosphere, hydrosphere and the geosphere. The chapter closely examines these key spheres of the Earth to provide an extensive understanding of the subject.

Everything in Earth's system can be placed into one of four major subsystems: land, water, living things, or air. These four subsystems are called "spheres." Specifically, they are the lithosphere (land), hydrosphere (water), biosphere (living things), and atmosphere (air). Each of these four spheres can be further divided into sub-spheres. To keep things simple in this course, there will be no distinction among the sub-spheres of any of the four major spheres.

Atmosphere

Atmosphere is the gas and aerosol envelope that extends from the ocean, land, and ice-covered surface of a planet outward into space. The density of the atmosphere decreases outward, because the gravitational attraction of the planet, which pulls the gases and aerosols (microscopic suspended particles of dust, soot, smoke, or chemicals) inward, is greatest close to the surface. Atmospheres of some planetary bodies, such as Mercury, are almost nonexistent, as the primordial atmosphere has escaped the relatively low gravitational attraction of the planet and has been released into space. Other planets, such as Venus, Earth, Mars, and the giant outer planets of the solar system, have retained an atmosphere. In addition, Earth's atmosphere has been able to contain water in each of its three phases (solid, liquid, and gas), which has been essential for the development of life on the planet.

The evolution of Earth's current atmosphere is not completely understood. It is thought that the current atmosphere resulted from a gradual release of gases both from the planet's interior and from the metabolic activities of life-forms—as opposed to the primordial atmosphere, which developed by outgassing (venting) during the original formation of the planet. Current volcanic gaseous emissions include water vapour (H_2O), carbon dioxide (CO_2), sulfur dioxide (SO_2), hydrogen sulfide (H_2S), carbon monoxide (CO), chlorine (Cl), fluorine (F), and diatomic nitrogen N_2; consisting of two atoms in a single molecule), as well as traces of other substances. Approximately 85 percent of volcanic emissions are in the form of water vapour. In contrast, carbon dioxide is about 10 percent of the effluent.

During the early evolution of the atmosphere on Earth, water must have been able to exist as a liquid, since the oceans have been present for at least three billion years. Given that solar output four billion years ago was only about 60 percent of what it is today, enhanced levels of carbon dioxide and perhaps ammonia (NH_3) must have been present in order to retard the loss of infrared radiation into space. The initial life-forms that evolved in this environment must have been anaerobic (i.e., surviving in the absence of oxygen). In addition, they must have been able to resist the biologically destructive ultraviolet radiation in sunlight, which was not absorbed by a layer of ozone as it is now.

Once organisms developed the capability for photosynthesis, oxygen was produced in large quantities. The buildup of oxygen in the atmosphere also permitted the development of the ozone layer as O_2 molecules were dissociated into monatomic oxygen (O; consisting of single oxygen atoms) and recombined with other O_2 molecules to form triatomic ozone molecules (O_3). The capability for photosynthesis arose in primitive forms of plants between two and three billion years ago. Previous to the evolution of photosynthetic organisms, oxygen was produced in limited quantities as a by-product of the decomposition of water vapour by ultraviolet radiation.

The current molecular composition of Earth's atmosphere is diatomic nitrogen (N_2), 78.08 percent; diatomic oxygen (O_2), 20.95 percent; argon (A), 0.93 percent; water (H_2O), about 0 to 4 percent; and carbon dioxide (CO_2), 0.04 percent. Inert gases such as neon (Ne), helium (He), and krypton (Kr) and other constituents such as nitrogen oxides, compounds of sulfur, and compounds of ozone are found in lesser amounts.

Surface Budgets

Energy Budget

Earth's atmosphere is bounded at the bottom by water and land—that is, by the surface of Earth. Heating of this surface is accomplished by three physical processes—radiation, conduction, and convection—and the temperature at the interface of the atmosphere and surface are a result of this heating.

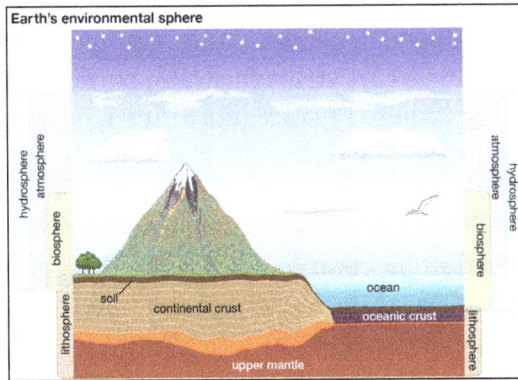

Earth's environmental spheres: Earth's environment includes the atmosphere, the hydrosphere, the lithosphere, and the biosphere.

The relative contributions of each process depend on the wind, temperature, and moisture structure in the atmosphere immediately above the surface, the intensity of solar insolation, and the physical characteristics of the surface. The temperature occurring at this interface is of critical importance in determining how suitable a location is for different forms of life.

Radiation

The temperature of the atmosphere and surface is influenced by electromagnetic radiation, and this radiation is traditionally divided into two types: insolation from the Sun and remittances from the surface and the atmosphere. Insolation is frequently referred to as shortwave radiation; it falls primarily within the ultraviolet and visible portions of the electromagnetic spectrum and consists predominantly of wavelengths of 0.39 to 0.76 micrometres (0.00002 to 0.00003 inch). Radiation emitted from Earth is called long wave radiation; it falls within the infrared portion of the spectrum and has typical wavelengths of 4 to 30 micrometres (0.0002 to 0.001 inch). Wavelengths of radiation emitted by a body depend on the temperature of the body, as specified by Planck's radiation law. The Sun, with its surface temperature of around 6,000 kelvins (K; about 5,725 °C, or 10,337 °F), emits at a much shorter wavelength than does Earth, which has lower surface and atmospheric temperatures around 250 to 300 K (−23 to 27 °C, or −9.4 to 80.6 °F).

A fraction of the incoming shortwave radiation is absorbed by atmospheric gases, including water vapour, and warms the air directly, but in the absence of clouds most of this energy reaches the surface. The scattering of a fraction of the shortwave radiation—particularly of the shortest wavelengths by air molecules in a process called Rayleigh scattering—produces Earth's blue skies.

When tall thick clouds are present, a large percentage (up to about 80 percent) of the insolation is reflected back into space. (The fraction of reflected shortwave radiation is called the cloud albedo.) Of the solar radiation reaching Earth's surface, some is reflected back into the atmosphere. Values of the surface albedo range as high as 0.95 for fresh snow to 0.10 for dark, organic soils. On land, this reflection occurs entirely at the surface. In water, however, albedo depends on the angle of the Sun's rays and the depth of the water column. If the Sun's rays strike the water surface at an oblique angle, albedo may be higher than 0.85; if these rays are more direct, only a small portion, perhaps as low as 0.02, is reflected, while the rest of the insolation is scattered within the water column and absorbed. Shortwave radiation penetrates a volume of water to significant depths (up to several hundred metres) before the insolation is completely attenuated. The heating by solar radiation in water is distributed through a depth, which results in smaller temperature changes at the surface of the water than would occur with the same insolation over an equal area of land.

The amount of solar radiation reaching the surface depends on latitude, time of year, time of day, and orientation of the land surface with respect to the Sun. In the Northern Hemisphere north of 23°30′, for example, solar insolation at local noon is less on slopes facing the north than on land oriented toward the south.

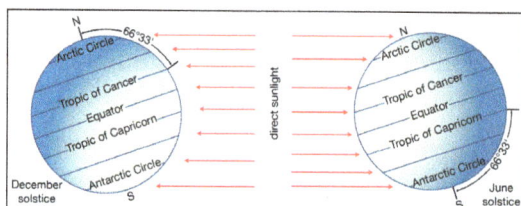

The primary cause of Earth's seasons is the change in the amount of sunlight reaching the surface at various latitudes over the course of a year. Because Earth is tilted on its axis with respect to the plane of its orbit around the Sun, different parts of its surface are in direct (overhead) sunlight at different times of the year.

Solar radiation is made up of direct and diffuse radiation. Direct shortwave radiation reaches the surface without being absorbed or scattered from its line of propagation by the intervening atmosphere. The image of the Sun's disk as a sharp and distinct object represents that portion of the solar radiation that reaches the viewer directly. Diffuse radiation, in contrast, reaches the surface after first being scattered from its line of propagation. On an overcast day, for example, the Sun's disk is not visible, and all of the shortwave radiation is diffuse.

Long-wave radiation is emitted by the atmosphere and propagates both upward and downward. According to the Stefan-Boltzmann law, the total amount of long-wave energy emitted is proportional to the fourth power of the temperature of the emitting material (e.g., the ground surface or the atmospheric layer). The magnitude of this radiation reaching the surface depends on the temperature at the height of emission and the amount of absorption that takes place between the height of emission and the surface. A larger fraction of the long-wave radiation is absorbed when the intervening atmosphere holds large amounts of water vapour and carbon dioxide. Clouds with liquid water concentrations near 2.5 grams per cubic metre absorb almost 100 percent of the long-wave radiation within a depth of 12 metres (40 feet) into the cloud. Clouds with lower liquid water concentrations require greater depths before complete absorption is attained (e.g., a cloud with a water content of 0.05 gram per cubic metre requires about 600 metres [about 2,000 feet] for complete absorption). Clouds that are at least this thick emit long-wave radiation from their bases downward to Earth's surface. The amount of long-wave radiation emitted corresponds to the temperature of the lowest levels of the cloud. (Clouds with warmer bases emit more long-wave radiation downward than colder clouds).

Conduction

The magnitude of heat flux by conduction below a surface depends on the thermal conductivity and the vertical gradient of temperature in the material beneath the surface. Soils such as dry peat, which has very low thermal conductivity (i.e., 0.06 watt per metre per K), permit little heat flux. In contrast, concrete has a thermal conductivity about 75 times as large (i.e., 4.60 watts per metre per K) and allows substantial heat flux. In water, the thermal conductivity is relatively unimportant, since, in contrast to land surfaces, insolation extends to substantial depths in the water; in addition, water can be mixed vertically.

Convection

Vertical mixing (convection) occurs in the atmosphere as well as in bodies of water. This process of mixing is also referred to as turbulence. It is a mechanism of heat flux that occurs in the atmosphere in two forms. When the surface is substantially warmer than the overlying air, mixing will spontaneously occur in order to redistribute the heat. This process, referred to as free convection, occurs when the environmental lapse rate (the rate of change of an atmospheric variable, such as temperature or density, with increasing altitude) of temperature decreases at a rate greater than 1 °C per 100 metres (approximately 1 °F per 150 feet). This rate is called the adiabatic lapse rate (the rate of temperature change occurring within a rising or descending air parcel). In the ocean, the temperature increase with depth that results in free convection is dependent on the temperature, salinity, and depth of the water. For example, if the surface has a temperature of 20 °C (68 °F) and a salinity of 34.85 parts per thousand, an increase in temperature with depth of greater than about 0.19 °C per km (0.55 °F per mile) just below in the upper layers of the ocean will result in free convection. In the atmosphere, the temperature profile with height determines whether free convection occurs or not. In the ocean, free convection depends on the temperature and salinity profile with depth. Colder and more saline conditions in a surface parcel of water, for example, make it more likely for that parcel to sink spontaneously and thus become part of the process of free convection.

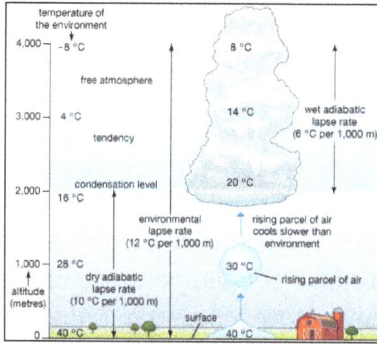

Rising air in an unstable atmosphere.

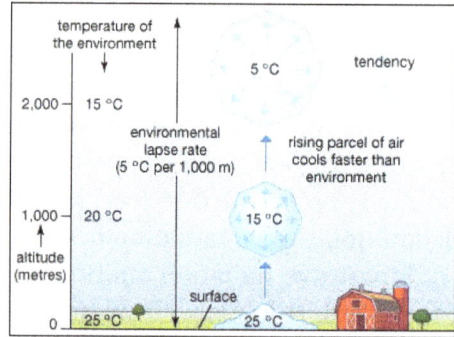

Rising air under stable conditions.

Mixing can also occur because of the shear stress of the wind on the surface. Shear stress is the pulling force of a fluid moving in one direction as it passes close to a fluid or object moving in another. As a result of surface friction, the average wind velocity at Earth's surface must be zero unless that surface is itself moving, such as in rivers or ocean currents. Winds above the surface decelerate when the vertical wind shear (the change in wind velocity at differing altitudes) becomes large enough to result in vertical mixing. The process by which heat and other atmospheric properties are mixed as a result of wind shear is called forced convection. Free and forced convection are also called convective and mechanical turbulence, respectively. This convection occurs as either sensible turbulent heat flux (heat directly transported to or from a surface) or latent turbulent heat flux (heat used to evaporate water from a surface). When this mixing does not occur, wind speeds are weak and change little with time; plumes from power-plant stacks within this layer, for example, spread very little in the vertical and remain in close proximity to the stacks.

Water Budget

The water budget at the air-surface interface is also of crucial importance in influencing atmospheric processes. The surface gains water through precipitation (rain and snow), direct condensation, and deposition (dew and frost). On land, the precipitation is often so large that some of it infiltrates into the ground or runs off into streams, rivers, lakes, and the oceans. Some of the precipitation remaining on the surface, such as in puddles or on vegetation, immediately evaporates back into the atmosphere.

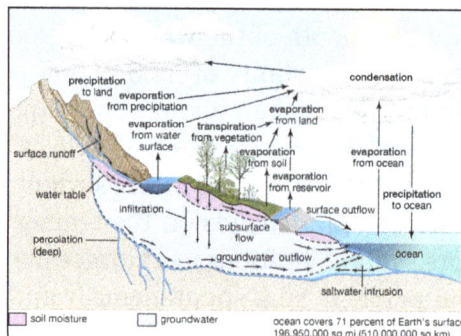

In the hydrologic cycle, water is transferred between the land surface, the ocean, and the atmosphere.

Liquid water in the soil is also converted to water vapour by transpiration from the leaves and stems of plants and by evaporation. The roots of vegetation may extract water from within the soil and emit it through stoma, or small openings, on the leaves. In addition, water may be evaporated from the surface of the soil directly, when groundwater from below is diffused upward. Evaporation occurs at the surface of water bodies at a rate that is inversely proportional to the relative humidity immediately above the surface. Evaporation is rapid in dry air but much slower when the lowest levels of the atmosphere are close to saturation. Evaporation from soils is dependent on the rate at which moisture is supplied by capillary suction within the soil, whereas transpiration from vegetation is dependent on both the water available within the root zone of plants and whether the stoma are open on the leaf surfaces. Water that evaporates and transpires into the atmosphere is often transported long distances before it precipitates out. The input, transport, and removal of water from the atmosphere is part of the hydrologic cycle. At any one time, only a very small fraction of Earth's water is present within the atmosphere; if all the atmospheric water was condensed out, it would cover the surface of the planet only to an average of about 2.5 cm (1 inch).

Nitrogen Budget

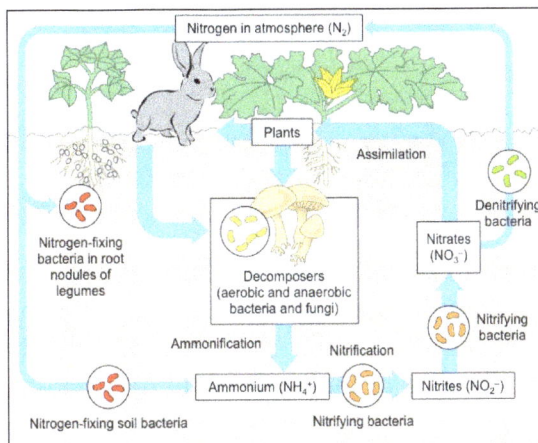

The nitrogen cycle

The nitrogen budget involves the chemical transformation of diatomic nitrogen (N_2), which makes up 78 percent of the atmospheric gases, into compounds containing ammonium (NH^+), nitrite (NO_2^-), and nitrate (NO_3^-). In a process called nitrification, or nitrogen fixation, bacteria such as *Rhizobium* living within nodules on the roots of peas, clover, and other legumes convert diatomic nitrogen gas to ammonia. A small amount of nitrogen is also fixed by lightning. Ammonia may be further transformed by other bacteria into nitrites and nitrates and used by plants for growth. These compounds are eventually converted back to N_2 after the plants die or are eaten by denitrifying bacteria. These bacteria, in their consumption of plants and both the excrement and corpses of plant-eating animals, convert much of the nitrogen compounds back to N_2. Some of these compounds are also converted to N_2 by a series of chemical processes associated with ultraviolet light from the Sun. The combustion of petroleum by motor vehicles also produces oxides of nitrogen, which enhance the natural concentrations of these compounds. Smog, which occurs in many urban areas, is associated with substantially higher levels of nitrogen oxides.

Sulfur Budget

The sulfur budget is also of major importance. Sulfur is put into the atmosphere as a result of weathering of sulfur-containing rocks and by intermittent volcanic emissions. Organic forms of sulfur are incorporated into living organisms and represent an important component in both the structure and the function of proteins. Sulfur also appears in the atmosphere as the gas sulfur dioxide (SO_2) and as part of particulate compounds containing sulfate (SO_4). Alone, both are directly dry-deposited or precipitated out onto Earth's surface. When wetted, these compounds are converted to caustic sulfuric acid (H_2SO_4).

Mount St. Helens volcano, viewed from the south during its eruption on May 18, 1980.

Since the beginning of the Industrial Revolution, human activities have injected significant quantities of sulfur into the atmosphere through the combustion of fossil fuels. In and near regions of urbanization and heavy industrial activity, the enhanced deposition and precipitation of sulfur in the form of sulfuric acid, and of nitrogen oxides in the form of nitric acid (HNO_3), resulting from vehicular emissions, have been associated with damage to fish populations, forests, statues, and building exteriors. The conversion of sulfur and nitrogen oxides to acids such as H_2SO_4 and HNO_3 is commonly known as the acid rain problem. Sulfur and nitrogen oxides are precipitated in rain, snow, and dry deposition (deposition to the surface during dry weather).

Carbon Budget

The carbon budget in the atmosphere is of critical importance to climate and to life. Carbon appears in Earth's atmosphere primarily as carbon dioxide (CO_2) produced naturally by the respiration of living organisms, the decay of these organisms, the weathering of carbon-containing rock strata, and volcanic emissions. Plants utilize CO_2, water, and solar insolation to convert CO_2 to diatomic oxygen (O_2). This process, known as photosynthesis, can result in local reductions of CO_2 of tens of parts per million within vegetation canopies. In contrast, night time respiration occurring when photosynthesis is not active can increase CO_2 concentrations. These concentrations may even double within dense tropical forest canopies for short periods before sunrise. On the global scale, seasonal variations of about 1 percent occur as a result of CO_2 uptake from photosynthesis, plant respiration, and soil respiration. Atmospheric CO_2 is primarily absorbed in the Northern Hemisphere during the growing season (spring to autumn). CO_2 is also absorbed by ocean waters; the rate of exchange to the ocean is greater for colder than for warmer waters. Currently CO_2 makes up about 0.03 percent of the gaseous composition of the atmosphere.

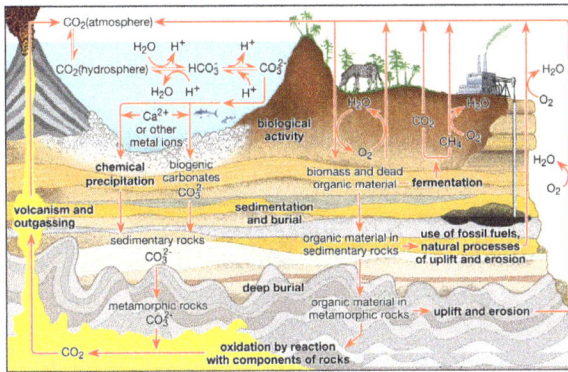

The carbon cycle.

Carbon is transported in various forms through the atmosphere, the hydrosphere, and geologic formations. One of the primary pathways for the exchange of carbon dioxide (CO_2) takes place between the atmosphere and the oceans; there a fraction of the CO_2 combines with water, forming carbonic acid (H_2CO_3) that subsequently loses hydrogen ions (H^+) to form bicarbonate (HCO_3^-) and carbonate (CO_3^{2-}) ions. Mollusk shells or mineral precipitates that form by the reaction of calcium or other metal ions with carbonate may become buried in geologic strata and eventually release CO_2 through volcanic outgassing. Carbon dioxide also exchanges through photosynthesis in plants and through respiration in animals. Dead and decaying organic matter may ferment and release CO_2 or methane (CH_4) or may be incorporated into sedimentary rock, where it is converted to fossil fuels. Burning of hydrocarbon fuels returns CO_2 and water (H_2O) to the atmosphere. The biological and anthropogenic pathways are much faster than the geochemical pathways and, consequently, have a greater impact on the composition and temperature of the atmosphere.

In the geologic past, CO_2 levels have been significantly higher than they are today and have had a significant effect on both climate and ecology. During the Carboniferous Period (360 to 300 million years ago), for example, moderately warm and humid climates combined with high concentrations of CO_2 were associated with extensive lush vegetation. After these plants died and decomposed, they were converted to sedimentary rocks that eventually became the coal deposits currently used for industrial combustion.

In the atmosphere, certain wavelengths of long-wave radiation are absorbed and then reemitted by CO_2. Since the lower levels of the atmosphere are warmer than layers higher up, the absorption of upward-propagating electromagnetic radiation, and a re-emission of a portion of it back downward, permits the lower atmosphere to remain warmer than it would be otherwise. The association of higher concentrations of CO_2 in the air with a warmer lower troposphere is commonly referred to as the greenhouse effect. (The name is inaccurate—an actual greenhouse is warmed primarily because solar radiation enters through the glass, which retains the heated air and prevents the mixing of cooler air into the greenhouse from above.) In recent years, there has been increasing concern that the release of CO_2 through the burning of coal and other fossil fuels will warm the lower atmosphere, a phenomenon commonly referred to as global warming. Water vapour is a more efficient greenhouse gas than carbon dioxide. However, since H_2O is ubiquitous, occurring in its three phases (solid, liquid, and gas), and since CO_2 is also a bio geochemically active gas, global temperature changes are both explained and predicted by changes in the atmospheric concentration of CO_2.

Vertical Structure of the Atmosphere

Earth's atmosphere is segmented into two major zones. The homosphere is the lower of the two and the location in which turbulent mixing dominates the molecular diffusion of gases. In this region, which occurs below 100 km (about 60 miles) or so, the composition of the atmosphere tends to be independent of height. Above 100 km, in the zone called the heterosphere, various atmospheric gases are separated by molecular mass, with the lighter gases being concentrated in the highest layers. Above 1,000 km (about 600 miles), helium and hydrogen are the dominant species. Diatomic nitrogen (N_2), a relatively heavy gas, drops off rapidly with height and exists in only trace amounts at 500 km (300 miles) and above. This decrease in the concentration of heavier gases with height is largest during periods of low Sun activity, when temperatures within the heterosphere are relatively low. The transition zone, located at a height of around 100 km between the homosphere and heterosphere, is called the turbopause.

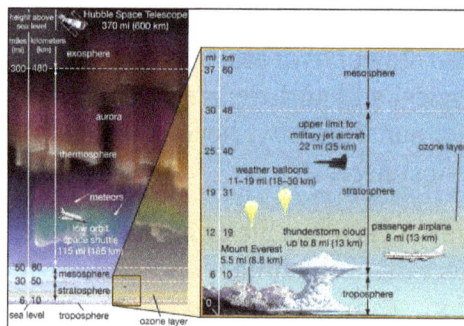

Atmosphere - vertical structure: The layers of Earth's atmosphere, showing heights of characteristic atmospheric phenomena.

The atmosphere can be further divided into several distinct layers defined by changes in air temperature with increasing height.

Troposphere

The lowest portion of the atmosphere is the troposphere, a layer where temperature generally decreases with height. This layer contains most of Earth's clouds and is the location where weather primarily occurs.

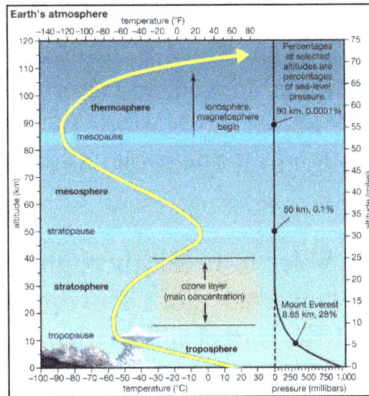

The layers of Earth's atmosphere: The yellow line shows the response of air temperature to increasing height.

Planetary Boundary Layer

The lower levels of the troposphere are usually strongly influenced by Earth's surface. This sublayer, known as the planetary boundary layer, is that region of the atmosphere in which the surface influences temperature, moisture, and wind velocity through the turbulent transfer of mass. As a result of surface friction, winds in the planetary boundary layer are usually weaker than above and tend to blow toward areas of low pressure. For this reason, the planetary boundary layer has also been called an Ekman layer, for Swedish oceanographer Vagn Walfrid Ekman, a pioneer in the study of the behaviour of wind-driven ocean currents.

Under clear, sunny skies over land, the planetary boundary layer tends to be relatively deep as a result of the heating of the ground by the Sun and the resultant generation of convective turbulence. During the summer, the planetary boundary layer can reach heights of 1 to 1.5 km (0.6 to 1 mile) above the land surface—for example, in the humid eastern United States—and up to 5 km (3 miles) in the south-western desert. Under these conditions, when unsaturated air rises and expands, the temperature decreases at the dry adiabatic lapse rate (9.8 °C per kilometre, or roughly 23 °F per mile) throughout most of the boundary layer. Near Earth's heated surface, air temperature decreases super adiabatically (at a lapse rate greater than the dry adiabatic lapse rate). In contrast, during clear, calm nights, turbulence tends to cease and radiational cooling (net loss of heat) from the surface results in an air temperature that increases with height above the surface.

When the rate of temperature decrease with height exceeds the adiabatic lapse rate for a region of the atmosphere, turbulence is generated. This is due to the convective overturn of the air as the warmer lower-level air rises and mixes with the cooler air aloft. In this situation, since the environmental lapse rate is greater than the adiabatic lapse rate, an ascending parcel of air remains warmer than the surrounding ambient air even though the parcel is both cooling and expanding. Evidence of this overturn is produced in the form of bubbles, or eddies, of warmer air. The larger bubbles often have sufficient buoyant energy to penetrate the top of the boundary layer. The subsequent rapid air displacement brings air from aloft into the boundary layer, thereby deepening the layer. Under these conditions of atmospheric instability, the air aloft cools according to the environmental lapse rate faster than the rising air is cooling at the adiabatic lapse rate. The air above the boundary layer replaces the rising air and undergoes compressional warming as it descends. As a result, this entrained air heats the boundary layer.

The ability of the convective bubbles to break through the top of the boundary layer depends on the environmental lapse rate aloft. The upward movement of penetrative bubbles will decrease rapidly if the parcel quickly becomes cooler than the ambient environment that surrounds it. In this situation, the air parcel will become less buoyant with additional ascent. The height that the boundary layer attains on a sunny day, therefore, is strongly influenced by the intensity of surface heating and the environmental lapse rate just above the boundary layer. The more rapidly a rising turbulent bubble cools above the boundary layer relative to the surrounding air, the lower the chance that subsequent turbulent bubbles will penetrate far above the boundary layer. The top of the daytime boundary layer is referred to as the mixed-layer inversion.

On clear, calm nights, radiational cooling results in a temperature increase with height. In this situation, known as a nocturnal inversion, turbulence is suppressed by the strong thermal stratification. Thermally stable conditions occur when warmer air overlies cooler, denser air. Over flat terrain, a nearly laminar wind flow (a pattern where winds from an upper layer easily slide past winds from a lower layer) can result. The depth of the radiationally cooled layer of air depends on a variety of factors, such as the moisture content of the air, soil and vegetation characteristics, and terrain configuration. In a desert environment, for instance, the nocturnal inversion tends to be found at greater heights than in a more humid environment. The inversion in more humid environments occurs at a lower altitude because more long-wave radiation emitted by the surface is absorbed by numerous available water molecules and reemitted back toward the surface. As a result, the lower levels of the troposphere are prevented from cooling rapidly. If the air is moist and sufficient near-surface cooling occurs, water vapour will condense into what is called "radiation fog."

Wind-generated Turbulence

During windy conditions, the mechanical production of turbulence becomes important. Turbulence eddies produced by wind shear tend to be smaller in size than the turbulence bubbles produced by the rapid convection of buoyant air. Within a few tens of metres of the surface during windy conditions, the wind speed increases dramatically with height. If the winds are sufficiently strong, the turbulence generated by wind shear can overshadow the resistance of layered, thermally stable air.

In general, there tends to be little turbulence above the boundary layer in the troposphere. Even so, there are two notable exceptions. First, turbulence is produced near jet streams, where large velocity shears exist both within and adjacent to cumuliform clouds. In these locations, buoyant turbulence occurs as a result of the release of latent heat. Second, pockets of buoyant turbulence may be found at and just above cloud tops. In these locations, the radiational cooling of the clouds destabilizes pockets of air and makes them more buoyant. Clear-air turbulence (CAT) is frequently reported when aircraft fly near one of these regions of turbulence generation.

The top of the troposphere, called the tropopause, corresponds to the level in which the pattern of decreasing temperature with height ceases. It is replaced by a layer that is essentially isothermal (of equal temperature). In the tropics and subtropics, the tropopause is high, often reaching to about 18 km (11 miles), as a result of vigorous vertical mixing of the lower atmosphere by thunderstorms. In polar regions, where such deep atmospheric turbulence is much less frequent, the tropopause is often as low as 8 km (5 miles). Temperatures at the tropopause range from as low as −80 °C (−112 °F) in the tropics to −50 °C (−58 °F) in polar region.

Cloud Formation within the Troposphere

The region above the planetary boundary layer is commonly known as the free atmosphere. Winds at this volume are not directly retarded by surface friction. Clouds occur most frequently in this portion of the troposphere, though fog and clouds that impinge or develop over elevated terrain often occur at lower levels.

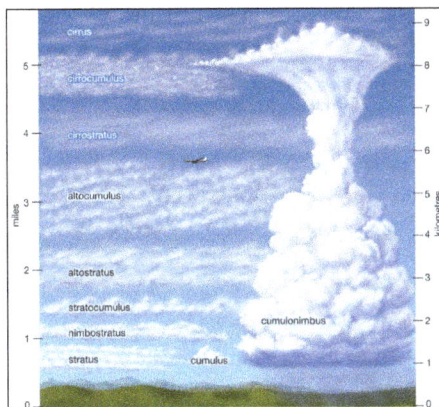

Cloud formation at various heights

There are two basic types of clouds: cumuliform and stratiform. Cloud types develop when clear air ascends, cooling adiabatically as it expands until either water begins to condense or deposition occurs. Water undergoes a change of state from gas to liquid under these conditions, because cooler air can hold less water vapour than warmer air.

For example, air at 20 °C (68 °F) can contain almost four times as much water vapour as at 0 °C (32 °F) before saturation takes place and water vapour condenses into liquid droplets.

Stratiform clouds occur as saturated air is mechanically forced upward and remains colder than the surrounding clear air at the same height. In the lower troposphere, such clouds are called stratus. Advection fog is a stratus cloud with a base lying at Earth's surface. In the middle troposphere, stratiform clouds are known as altostratus. In the upper troposphere, the terms cirrostratus and cirrus are used. The cirrus cloud type refers to thin, often wispy, cirrostratus clouds. Stratiform clouds that both extend through a large fraction of the troposphere and precipitate are called nimbostratus.

Cirrus fibratus are high clouds that are nearly straight or irregularly curved. They appear as fine white filaments and are generally distinct from one another.

Altocumulus radiatus, a cloud layer with laminae arranged in parallel bands.

Cumuliform clouds occur when saturated air is turbulent. Such clouds, with their bubbly turreted shapes, exhibit the small-scale up-and-down behaviour of air in the turbulent planetary boundary layer. Often such clouds are seen with bases at or near the top of the boundary layer as turbulent eddies generated near Earth's surface reach high enough for condensation to occur.

Cumulus humilis, flattened clouds characterized by only a small vertical extent.

Cumuliform clouds will form in the free atmosphere if a parcel of air, upon saturation, is warmer than the surrounding ambient atmosphere. Since this air parcel is warmer than its surroundings, it will accelerate upward, creating the saturated turbulent bubble characteristic of a cumuliform cloud. Cumuliform clouds, which reach no higher than the lower troposphere, are known as cumulus humulus when they are randomly distributed and as stratocumulus when they are organized into lines. Cumulus congestus clouds extend into the middle troposphere, while deep, precipitating cumuliform clouds that extend throughout the troposphere are called cumulonimbus. Cumulonimbus clouds are also called thunderstorms, since they usually have lightning and thunder associated with them. Cumulonimbus clouds develop from cumulus humulus and cumulus congestus clouds.

Stratosphere and Mesosphere

The stratosphere is located above the troposphere and extends up to about 50 km (30 miles). Above the tropopause and the isothermal layer in the lower stratosphere, temperature increases with height. Temperatures as high as 0 °C (32 °F) are observed near the top of the stratosphere. The observed increase of temperature with height in the stratosphere results in strong thermodynamic stability with little turbulence and vertical mixing. The warm temperatures and very dry air result in an almost cloud-free volume. The infrequent clouds that do occur are called nacreous, or mother-of-pearl, clouds because of their striking iridescence, and they appear to be composed of both ice and super cooled water. These clouds form up to heights of 30 km (19 miles).

The pattern of temperature increase with height in the stratosphere is the result of solar heating as ultraviolet radiation in the wavelength range of 0.200 to 0.242 micrometre dissociates diatomic oxygen (O_2). The resultant attachment of single oxygen atoms to O_2 produces ozone (O_3). Natural stratospheric ozone is produced mainly in the tropical and middle latitudes. Regions of nearly complete ozone depletion, which have occurred in the Antarctic during the spring, are associated with nacreous clouds, chlorofluorocarbons (CFCs), and other pollutants from human activities. These regions are more commonly known as ozone holes. Ozone is also transported downward into the troposphere, primarily in the vicinity of the polar front.

The stratopause caps the top of the stratosphere, separating it from the mesosphere near 45–50 km (28–31 miles) in altitude and a pressure of 1 millibar (approximately equal to 0.75 mm of mercury at 0 °C, or 0.03 inch of mercury at 32 °F). In the mesosphere, temperatures again decrease with increasing altitude. Unlike the situation in the stratosphere, vertical air currents in the mesosphere are not strongly inhibited. Ice crystal clouds, called noctilucent clouds, occasionally form in the upper mesosphere. Above the mesopause, a region occurring at altitudes near 85 to 90 km (50 to 55 miles), temperature again increases with height in a layer called the thermosphere.

Thermosphere

Temperatures in the thermosphere range from near 500 K (approximately 227 °C, or 440 °F) during periods of low sunspot activity to 2,000 K (1,725 °C, or 3,137 °F) when the Sun is active. The thermopause, defined as the level of transition to a more or less isothermal temperature profile at the top of the thermosphere, occurs at heights of around 250 km (150 miles) during quiet Sun periods and almost 500 km (300 miles) when the Sun is active. Above 500 km, molecular collisions are infrequent enough that temperature is difficult to define.

The portion of the thermosphere where charged particles (ions) are abundant is called the ionosphere. These ions result from the removal of electrons from atmospheric gases by solar ultraviolet radiation. Extending from about 80 to 300 km (about 50 to 185 miles) in altitude, the ionosphere is an electrically conducting region capable of reflecting radio signals back to Earth.

Maximum ion density, a condition that makes for efficient radio transmission, occurs within two sub layers the lower E region, which exists from 90 to 120 km (about 55 to 75 miles) in altitude; and the F region, which exists from 150 to 300 km (about 90 to 185 miles) in altitude. The F region has two maxima (i.e., two periods of highest ion density) during daylight hours, called F1 and F2. Both the F1 and F2 regions possess high ion density and are strongly influenced by both solar activity and time of day. Of these, the F2 region is the more variable of the two and may reach an ion density as high as 106 electrons per cubic centimetre. Shortwave radio transmissions, capable of reaching around the world, take advantage of the ability of layers in the ionosphere to reflect certain wavelengths of electromagnetic radiation. In addition, electrical discharges from the tops of thunderstorms into the ionosphere, called transient luminous events, have been observed.

Magnetosphere and Exosphere

Above approximately 500 km (300 miles), the motion of ions is strongly constrained by the presence of Earth's magnetic field. This region of Earth's atmosphere, called the magnetosphere, is compressed by the solar wind on the daylight side of the planet and stretched outward in a long tail on the night side. The colourful auroral displays often seen in polar latitudes are associated with bursts of high-energy particles generated by the Sun. When these particles are influenced by the magnetosphere, some are subsequently injected into the lower ionosphere.

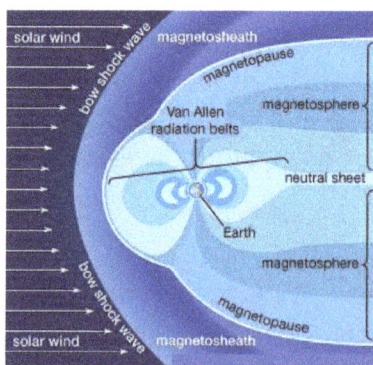

The Van Allen radiation belts contained within Earth's magnetosphere. Pressure from the Solar wind is responsible for the asymmetrical shape of the magnetosphere and the belts.

The layer above 500 km is referred to as the exosphere, a region in which at least half of the upward-moving molecules do not collide with one another. In contrast, these molecules follow long ballistic trajectories and may exit the atmosphere completely if their escape velocities are high enough. The loss rate of molecules through the exosphere is critical in determining whether Earth or any other planetary body retains an atmosphere.

Horizontal Structure of the Atmosphere

Distribution of Heat From the Sun

The primary driving force for the horizontal structure of Earth's atmosphere is the amount and distribution of solar radiation that comes in contact with the planet. Earth's orbit around the Sun is an ellipse, with a perihelion (closest approach) of 147.5 million km (91.7 million miles) in early January and an aphelion (farthest distance) of 152.6 million km (94.8 million miles) in early July. As a result of Earth's elliptical orbit, the time between the autumnal equinox and the following vernal equinox is almost one week shorter than the remainder of the year in the Northern Hemisphere. This results in a shorter astronomical winter in the Northern Hemisphere than in the Southern Hemisphere.

Earth rotates once every 24 hours around an axis that is tilted at an angle of 23°30′ with respect to the plane of its orbit around the Sun. As a result of this tilt, during the summer season of either the Northern or the Southern Hemisphere, the Sun's rays are more direct at given latitude than they are during the winter season. Poleward of latitudes 66°30′ N and 66°30′ S, the tilt of the planet is such that for at least one complete day (at 66°30′) and as long as six months (at 90°), the Sun is above the horizon during the summer season and below the horizon during the winter.

As a result of this asymmetric distribution of solar heating, during the winter season the troposphere in the high latitudes becomes very cold. In contrast, during the summer at high latitudes, the troposphere warms significantly as a result of the long hours of daylight; however, owing to the oblique angle of the sunlight near the poles, the temperatures there remain relatively cool compared with middle latitudes. Equator ward of latitudes 30° N and 30° S or so, substantial radiant heating from the Sun occurs during both winter and summer seasons. The tropical troposphere, therefore, has comparatively little variation in temperature during the year.

Convection, Circulation and Deflection of Air

The region of greatest solar heating at the surface in the humid tropics corresponds to areas of deep cumulonimbus convection. Cumulonimbus clouds routinely form in the tropics where rising parcels of air are warmer than the surrounding ambient atmosphere. They transport water vapour, sensible heat, and Earth's rotational momentum to the upper portion of the troposphere. As a result of the vigorous convective mixing of the atmosphere, the tropopause in the lower latitudes is often very high, located some 17 to 18 km (10.5 to 11 miles) above the surface.

Since motion upward into the stratosphere is inhibited by very stable thermal layering, the air transported upward by convection diverges toward the poles in the upper troposphere. (This divergence aloft results in a wide strip of low atmospheric pressure at the surface in the tropics, occurring in an area called the equatorial trough). As the diverted air in the troposphere moves toward the poles, it tends to retain the angular momentum of the near-equatorial region, which is large as a result of Earth's rotation. As a result, the poleward-moving air is deflected toward the right in the Northern Hemisphere and toward the left in the Southern Hemisphere.

Upon reaching around 30° of latitude poleward of its region of origin, the upper-level air is traveling primarily toward the poles and is tending toward the east. Since motion upward is constrained by the stratosphere, the slowly cooling air must descend. The compressional warming that occurs as the air descends creates vast regions of subtropical high pressure. These regions are centred over the oceans and are characterized by strong thermodynamic stability. The sparse precipitation in these regions, a result of stability and subsidence, is associated with such great arid regions of the world as the Sahara, Atacama, Kalahari, and Sonoran deserts. The accumulation of air as a result of the convergence in the upper troposphere causes deep high-pressure systems, known as subtropical ridges, to form in these regions. Locally, these ridges are given such names as the Bermuda High, the Azores High, and the North Pacific High.

The Erg Admer, a large area of sand dunes in southern Algeria, is located within a vast region of subtropical high pressure. The arid conditions here result from the constant presence of descending air containing little moisture.

The descending air referred to above, upon reaching the lower troposphere, is forced to diverge by the presence of Earth's surface. Some air moves poleward, while the remainder moves equatorward. In either direction, the air is deflected to the right in the Northern Hemisphere and to the left in the Southern Hemisphere. Deflection occurs because, in accordance with Newton's first law of motion, a parcel moving in a certain direction will retain the same motion unless acted on by an exterior force. With respect to a rotating Earth, a moving parcel conserving its momentum (i.e., not acted on by an exterior force) will appear to be deflected with respect to fixed points on the rotating

Earth. As seen from a fixed point in space, such a parcel would be moving in a straight line. This apparent force on the motion of a fluid (in this case, air) is called the Coriolis effect. As a result of the Coriolis effect, air tends to rotate counterclockwise around large-scale low-pressure systems and clockwise around large-scale high-pressure systems in the Northern Hemisphere. In the Southern Hemisphere, the flow direction is reversed.

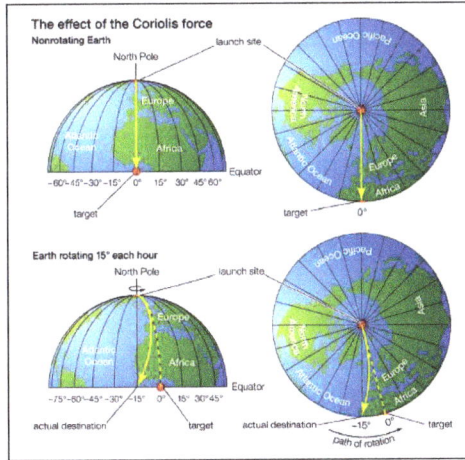

The effect of the Coriolis force

In the equatorward-moving flow, this deflection results in northeast winds north of 0° latitude and southeast winds south of that latitude. These low-level winds have been called the trade winds since 17th-century sailing vessels used them to travel to the Americas. The convergence region for lower-level northeast and southeast trade winds is called the intertropical convergence zone (ITCZ). The ITCZ corresponds to the equatorial trough and is the mechanism that helps generate the deep cumulonimbus clouds through convection. Cumulonimbus clouds are the main conduit transporting tropical heating into the upper troposphere.

The circulation pattern described above—ascent in the equatorial trough, poleward movement in the upper troposphere, descent in the subtropical ridges, and equatorward movement in the trade winds—is in effect a direct heat engine, which meteorologists call the Hadley cell. This persistent circulation mechanism transports heat from the latitudes of greatest solar insolation to the latitudes of the subtropical ridges. The geographic location of the Hadley circulation moves north and south with the seasons; however, the equatorial trough lags behind for about two months owing to the thermal inertia of Earth's surface. (For a given location on Earth's surface, the highest daily temperatures are achieved just after the period of greatest insolation, since time is required to heat the ocean surface waters and the soil).

Extratropical Cyclones

Poleward of the subtropical ridges, winds in the lower troposphere tend to be southwesterly in the Northern Hemisphere and northwesterly in the Southern Hemisphere, again owing to the Coriolis effect. Since warm air is being moved poleward at low altitudes, the wind flow is no longer associated with the direct heat engine of the Hadley cell. Instead, the continued transport of heat from the equatorial trough toward the poles is facilitated by large low-pressure eddies called extratropical cyclones. These phenomena develop along the polar front, which separates colder polar air from warmer tropical air, when sufficiently large temperature differences occur across the frontal boundary in the lower troposphere. The intensity of this temperature gradient is referred to as the baroclinicity of the front.

Extratropical cyclones have three stages of expansion: the developing stage, in which an undulating wave develops along the front; the mature stage, in which sinking cold air sweeps equatorward west of the surface low-pressure centre and ascending warm air moves poleward east of the cyclone; and the occluded stage, in which the warm air is entrained within and moved above the polar air and becomes separated from the source region of the tropical air. Cyclones that progress no farther than the developing stage are referred to as wave cyclones, while extratropical lows that reach the mature and occluded stages are called baroclinically unstable waves. Extratropical storm development is referred to as cyclogenesis. Rapid extratropical cyclone development, called explosive cyclogenesis, is often associated with major winter storms and occurs when surface pressure falls by more than about 24 millibars per day. Theoretical analysis has shown that the occurrence of baroclinically unstable waves is directly proportional to the magnitude of the temperature gradient, with maximum growth for wavelengths of 3,000 to 5,000 km (1,865 to 3,100 miles). Wavelengths that are shorter are damped by horizontal mixing. The 3,000 to 5,000 km wavelength is the typical separation between high- and low- pressure synoptic weather systems in the middle and higher latitudes.

Polar Fronts and Jet Stream

In the troposphere, the demarcation between polar air and warmer tropical atmosphere is usually defined by the polar front. On the poleward side of the front, the air is cold and denser; equatorward of the front, the air is warmer and more buoyant. During the winter season, the polar front is generally located at lower latitudes and is more pronounced than in the summer.

Cold fronts occur at the leading edge of equatorward-moving polar air. In contrast, warm fronts are well defined at the equatorward surface position of polar air as it retreats on the eastern sides of extratropical cyclones. Equatorward-moving air behind a cold front occurs in pools of dense high pressure known as polar highs and arctic highs. The term arctic high is used to define air that originates even deeper within the high latitudes than polar highs.

When polar air neither retreats nor advances, the polar front is called a stationary front. In the occluded stage of the life cycle of an extratropical cyclone, when cold air west of the surface low-pressure centre advances more rapidly toward the east than cold air ahead of the warm front, warmer, less-dense air is forced aloft. This frontal intersection is called an occluded front. Without exception, fronts of all types follow the movement of colder air.

Clouds and often precipitation occur on the poleward sides of both warm and stationary fronts and whenever tropical air reaching the latitude of the polar front is forced upward over the colder air near the surface. Such fronts are defined as active fronts. Rain and snowfall from active fronts form a major part of the precipitation received in the middle and high latitudes. Precipitation in these areas occurs primarily during the winter months.

The position of the polar front slopes upward toward colder air. This occurs because cold air tends to undercut the warmer air of tropical origin. Since cold air is denser, atmospheric pressure decreases more rapidly with height on the poleward side of the polar front than on the warmer tropical side. This creates a large horizontal temperature contrast, which is essentially a large pressure gradient, between the polar and tropical air. In the middle and upper parts of the troposphere, this pressure gradient is responsible for the strong westerly winds occurring there. Winds created aloft circulate around a large region of upper-level low pressure near each of the poles. The centre of each low pressure region is a persistent cyclone known as the circumpolar vortex.

The region of strongest winds, which occurs at the juncture of the tropical and polar air masses, is called the jet stream. Since the temperature contrast between the tropics and the high latitudes is greatest in the winter, the jet stream is stronger during that season. In addition, since the mid-latitudes also become colder during the winter, while tropical temperatures remain relatively unchanged, the westerly jet stream approaches latitudes of 30° during the colder season. During the warmer season in both hemispheres, the jet stream moves poleward and is located between latitudes of 50° and 60°.

The jet stream reaches its greatest velocity at the tropopause. Above that level, a reversal of the horizontal temperature gradient occurs, which produces a reduction in the wind speeds of the jet stream at high latitudes. This causes a weakening of the westerlies with increasing height. At intervals ranging from 20 to 40 months, with a mean value of 26 months, westerly winds in the stratosphere reverse direction over low latitudes, so that an easterly flow develops. This feature is called the quasi-biennial oscillation (QBO). In addition, a phenomenon called sudden stratospheric warming, apparently the result of strong downward air motion, also occurs in the late winter and spring at high latitudes. Sudden stratospheric warming can significantly alter temperature-dependent chemical reactions of ozone and other reactive gases in the stratosphere and affect the development of such features as "ozone holes."

A major focus of weather forecasting in the middle and high latitudes is to forecast the movement and development of extratropical cyclones, polar and arctic highs, and the location and intensity of subtropical ridges. Spring and fall frosts, for example, are associated with the equatorward movement of polar highs behind a cold front, while droughts and heat waves in the summer are associated with unusually strong subtropical ridges.

Effect of Continents on Air Movement

Preferred geographic locations exist for subtropical ridges and for the development, movement, and decay of extratropical cyclones. During the winter months in middle and high latitudes, the lower parts of the troposphere over continents often serve as reservoirs of cold air as heat is radiated into space throughout the long nights. In contrast, the oceans lose heat less rapidly, because of the large heat capacity of water, their ability to overturn as the surfaces cool and become negatively buoyant, and the movement of ocean currents such as the Gulf Stream and the Kuroshio current. Warm currents transport heat from lower latitudes poleward and tend to occur on the western sides of oceans. The lower troposphere over these warmer oceanic areas tends to be a region of relative low pressure. As a result of this juxtaposition of cold air and warm air, the eastern sides of continents and the western fringes of oceans in middle and high latitudes are the preferred locations for extratropical storm development. Over Asia in particular, the cold high-pressure system is sufficiently permanent that a persistent offshore flow called the winter monsoon occurs.

An inverse type of flow develops in the summer as the continents heat more rapidly than their adjacent oceanic areas. Continental areas tend to become regions of relative low pressure, while high pressure in the lower troposphere becomes more prevalent offshore. As the winds travel from areas of higher pressure to areas of lower pressure, a persistent onshore flow develops over large landmasses in the lower troposphere. The result of this heating is referred to as the summer monsoon. The leading edge of this monsoon is associated with a feature called the monsoon trough, a region of low atmospheric pressure at sea level.

As a result of the continental effect, the subtropical ridge is segmented into surface high-pressure cells. In the summer, large landmasses in the subtropics tend to be centres of relative low pressure as a result of strong solar heating. As a consequence, persistent high-pressure cells, such as the Bermuda and Azores highs, occur over the oceans. The oval shape of these high-pressure cells creates a thermal structure on their eastern sides that differs from the thermal structure on their western sides in the lower troposphere. On the eastern side, subsidence from the Hadley circulation is enhanced by the tendency of air to preserve its angular momentum on the rotating Earth. Owing to the enhanced descent of air over the eastern parts of the oceans, landmasses adjacent to these areas (typically the western sides of continents) tend to be deserts, such as those found in northwestern and southwestern Africa and along western coastal Mexico.

Effect of Oceans on Air Movement

The arid conditions found along the western coasts of continents in subtropical latitudes are further enhanced by the influence of the equatorward surface air flow on the
ocean currents. This flow exerts a shearing stress on the ocean surface, which results
in the deflection of the upper layer of water above the thermocline to the right in the
Northern Hemisphere and to the left in the Southern Hemisphere. (This deflection is
also the result of the Coriolis effect; water from both hemispheres moves westward
when displaced toward the Equator.) As warmer surface waters are carried away by
this offshore ocean airflow, cold water from below the thermocline rises to the surface
in a process called upwelling. Upwelling creates areas of cold coastal surface waters
that stabilize the lower troposphere and reduce the chances for convection. Lower convection in turn reduces the likelihood for precipitation, although fogs and low stratus
clouds are common. Upwelling regions are also associated with enriched sea life, as oxygen and organic nutrients are transported upward from the depths toward the surface
of the ocean.

During periods when the intertropical convergence zone (ITCZ) is located near the
Equator, trade winds from the northeast and southeast converge there. The west-
ward-moving winds cause the displacement of surface ocean waters away from the
Equator such that the deeper, colder waters move to the surface. In the central and
eastern Pacific Ocean near the Equator, when this upwelling is stronger than average,
the event is called La Niña. When the trade winds weaken in this region, however,
warmer-than-average surface conditions occur, and upwelling is weaker than usual.
This event is called El Niño. Changes in ocean surface temperatures caused by El Niño
significantly affect where cumulonimbus clouds form in the ITCZ and, therefore, the
geographic structure of the Hadley cell. During periods when El Niño is active, weather
patterns across the entire Earth are substantially altered.

Mountain Barriers

North-south-oriented mountain barriers, such as the Rockies and the Andes, and large
massifs, such as the Plateau of Tibet, also influence atmospheric flow. When the general
westerly flow in the mid-latitudes reaches these barriers, air tends to be blocked. It is
transported poleward west of the terrain and toward the Equator east of the obstacle.
Air forced up the slopes of mountain barriers is often sufficiently moist to produce considerable precipitation on windward sides of mountains, whereas subsiding air on the
lee slopes produces more-arid conditions. Essentially, the elevated terrain affects the atmosphere as if it were an anticyclone, a centre of high pressure. In addition, mountains
prevent cold air from the continental interior from moving westward of the terrain. As a
result, relatively mild weather occurs along the western coasts of continents with north-
south mountain ranges when compared with continental interiors. For example, the
West Coast of North America experiences milder winter weather than the Great Plains
and Midwest, both of which occur at similar latitudes. In contrast, east-west mountain

barriers, such as the Alps in Europe, offer little impediment to the general westerly flow of air. In these situations, milder maritime conditions extend much farther inland.

Cloud Processes

Condensation

The formation of cloud droplets and cloud ice crystals is associated with suspended aerosols, which are produced by natural processes as well as human activities and are ubiquitous in Earth's atmosphere. In the absence of such aerosols, the spontaneous conversion of water vapour into liquid water or ice crystals requires conditions with relative humidities much greater than 100 percent, with respect to a flat surface of H_2O. The development of clouds in such a fashion, which occurs only in a controlled laboratory environment, is referred to as homogeneous nucleation. Air containing water vapour with a relative humidity greater than 100 percent, with respect to a flat surface, is referred to as being supersaturated. In the atmosphere, aerosols serve as initiation sites for the condensation or deposition of water vapour. Since their surfaces are of discrete sizes, aerosols reduce the amount of supersaturation required for water vapour to change its phase and are referred to as cloud condensation nuclei.

The larger the aerosol and the greater its solubility, the lower the supersaturation percentage required for the aerosol to serve as a condensation surface. Condensation nuclei in the atmosphere become effective at supersaturations of around 0.1 to 1 percent (that is, levels of water vapour around 0.1 to 1 percent above the point of saturation). The concentration of cloud condensation nuclei in the lower troposphere at a supersaturation of 1 percent ranges from around 100 per cubic centimetre (approximately 1,600 per cubic inch) in size in oceanic air to 500 per cubic centimetre (8,000 per cubic inch) in the atmosphere over a continent. Higher concentrations occur in polluted air.

Aerosols that are effective for the conversion of water vapour to ice crystals are referred to as ice nuclei. In contrast to cloud condensation nuclei, the most effective ice nuclei are hydrophobic (having a low affinity for water) with molecular spacings and a crystallographic structure close to that of ice.

While cloud condensation nuclei are always readily available in the atmosphere, ice nuclei are often scarce. As a result, liquid water cooled below 0 °C (32 °F) can often remain liquid at subfreezing temperatures because of the absence of effective ice nuclei. Liquid water at temperatures less than 0 °C is referred to as supercooled water. Except for true ice crystals, which are effective at 0 °C, all other ice nuclei become effective at temperatures below freezing. In the absence of any ice nuclei, the freezing of supercooled water droplets of a few micrometres in radius, in a process called homogeneous ice nucleation, requires temperatures at or lower than −39 °C (−38 °F). While a raindrop will freeze near 0 °C, small cloud droplets have too few molecules to create an ice crystal by random chance until the molecular motion is slowed as the temperature approaches −39 °C. When ice nuclei are present, heterogeneous ice nucleation can occur at warmer temperatures.

Ice nuclei are of three types: deposition nuclei, contact nuclei, and freezing nuclei. Deposition nuclei are analogous to condensation nuclei in that water vapour directly deposits as ice crystals on the aerosol. Contact and freezing nuclei, in contrast, are associated with the conversion of supercooled water to ice. A contact nucleus converts liquid water to ice by touching a supercooled water droplet. Freezing nuclei are absorbed into the liquid water and convert the supercooled water to ice from the inside out.

Examples of cloud condensation nuclei include sodium chloride (NaCl) and ammonium sulfate ($[NH_4]_2 SO_2$), whereas the clay mineral kaolinite is an example of an ice nuclei. In addition, naturally occurring bacteria found in decayed leaf litter can serve as ice nuclei at temperatures of less than about −4 °C (24.8 °F). In a process called cloud seeding, silver iodide, with effective ice-nucleating temperatures of less than −4 °C, has been used for years in attempts to convert supercooled water to ice crystals in regions with a scarcity of natural ice nuclei.

Precipitation

Liquid Droplets

The evolution of clouds that follows the formation of liquid cloud droplets or ice crystals depends on which phase of water occurs. A cloud in which only liquid water occurs (even at temperatures less than 0 °C) is referred to as a warm cloud, and the precipitation that results is said to be due to warm-cloud processes. In such a cloud, the growth of a liquid water droplet to a raindrop begins with condensation, as additional water vapour condenses in a supersaturated atmosphere. This process continues until the droplet has attained a radius of about 10 micrometres (0.0004 inch). Above this size, since the mass of the droplet increases according to the cube of its radius, further increases by condensational growth are very slow. Subsequent growth, therefore, occurs only when the cloud droplets develop at slightly different rates. Differences in growth rates have been attributed to differences in spatial variations of the initial aerosol sizes, in solubility's, and in magnitudes of super saturation. Cloud droplets of different sizes will fall at different velocities and will collide with droplets of different radii. If the collision is hard enough to overcome the surface tension between the two colliding droplets, coalescence will occur and result in a new and larger single droplet.

This process of cloud-droplet growth is referred to as collision-coalescence. Warm-cloud rain results when the droplets attain a sufficient size to fall to the ground. Such a raindrop (perhaps about 1 mm [0.04 inch] in radius) contains perhaps one million 10-micrometre cloud droplets. The typical radii of raindrops resulting from this type of precipitation process range up to several millimetres and have fall velocities of around 3 to 4 metres (10 to 13 feet) per second. This type of precipitation is very common from shallow cumulus clouds over tropical oceans. In these locations, the concentration of cloud condensation nuclei is so small that there is only limited competition for the available water vapour.

Precipitation of Ice

A cloud that contains ice crystals is referred to as a cold cloud, and the resulting precipitation is said to be the product of cold-cloud processes. Traditionally, this process has also been referred to as the Bergeron-Findeisen mechanism, for Swedish meteorologists Tor Bergeron and Walter Findeisen, who introduced it in the 1930s. In this type of cloud, ice crystals can grow directly from the deposition of water vapour. This water vapour may be supersaturated with respect to ice, or it may be the result of evaporation of supercooled water and subsequent deposition onto an ice crystal. Since the saturation vapour pressure of liquid water is always greater than or equal to the saturation vapour pressure of ice, ice crystals will grow at the expense of the liquid water. For example, saturated air with respect to liquid water becomes supersaturated with respect to ice by 10 percent at −10 °C (14 °F) and by 21 percent at −20 °C (-4 °F). This results in a rapid conversion of liquid water to ice. This substantial and rapid change of phase permits large ice crystals in a cloud surrounded by a large number of supercooled cloud droplets to grow quickly (often in less than 15 minutes) from tiny ice crystals to snowflakes. These snowflakes are large enough to fall by depositional growth alone. Fall velocities of snow range up to about 2 metres per second (6.5 feet per second). Ice crystals that grow by deposition have much lower densities than solid ice because of the air pockets occurring within the volume of the crystal. This lower density differentiates snow from ice. Clouds that are completely converted to ice crystals are referred to as glaciated clouds.

The specific form the ice crystals take depends on the temperature and the degree of supersaturation with respect to ice. At −14 °C (7 °F) and a relatively large supersaturation with respect to liquid water, for example, ice crystals with dendritic (treelike branching) patterns form. This type of ice crystal, the one usually used to represent snowflakes in photographs and drawings, experiences growth at the end of radial arms on one or more planes of the crystal. At −40 °C (−40 °F) and a supersaturation with respect to liquid water of close to 0 percent, hollow ice columns form.

Ice crystals can also grow large enough to precipitate either by aggregation or by riming. Aggregation occurs when the arms of the ice crystals interlock and form a clump. This collection of intermingled ice crystals can occasionally reach several centimetres in diameter. Ice crystals can also grow when supercooled water freezes directly onto the crystal to form rime. With greater accumulation of dense ice on the crystal, its fall velocity increases. When the riming is substantial enough, the crystal form of the snowflake is lost and replaced by a more or less spherical particle called graupel. Smaller-sized graupels are generally referred to as snow grains. In cumulonimbus clouds during conditions where graupels are repeatedly wetted and then injected back toward high altitudes by strong updrafts, very large graupels called hail result. Hail has been observed on the ground at sizes larger than grapefruits.

Frozen precipitation, falling to levels of the atmosphere that are much warmer than 0 °C, often melts and reaches the ground as rain. Such cold-cloud rain at the ground is usually distinguished from warm-cloud rain by its larger size. Melted hailstones, in particular, make a large-radius impact when they strike the ground. Cold-cloud rain occasionally will refreeze if a layer of subfreezing air exists near Earth's surface. When this freezing occurs in the free atmosphere, the frozen raindrops are referred to as sleet or ice pellets. When this freezing occurs only upon the impact of the raindrop with the ground, the precipitation is known as freezing rain. During ice storms, freezing rain can produce accumulations heavy enough to snap large trees and electrical lines.

Lightning and Optical Phenomena

The repeated collision of ice crystals and graupel in clouds is associated with the buildup of electrical charge. This electrification is particularly large in cumulonimbus clouds as a result of vigorous vertical mixing and collisions. On average, positive charges accumulate in the upper regions, while negative charges are concentrated lower down. In response to the negative charge near the cloud base, and as negatively charged rain falls toward the ground, a pocket of positive charge develops on the ground. When the difference in electric potential between positive and negative charges becomes large enough, a sudden electrical discharge (lightning) will occur. Lightning can occur between different regions of the cloud, as in intracloud lightning, and between the cloud and the positively charged ground, as in cloud-to-ground lightning. The passage of the lightning through the air heats it to above 30,000 K (29,725 °C, or 53,540 °F), causing a large increase in pressure. This produces a powerful shock wave that is heard as thunder.

Sunlight that propagates through clouds and precipitation often produces fascinating optical images. Rainbows are produced when sunlight is diffracted into its component colours by water droplets. In addition, halos are produced by the refraction and reflection of sunlight or moonlight by ice crystals, while coronas are formed when sunlight or moonlight passes through water droplets.

Sunlight: Rays of sunlight shining through clouds.

Biosphere

Biosphere is relatively thin life-supporting stratum of Earth's surface, extending from a few kilometres into the atmosphere to the deep-sea vents of the ocean. The biosphere is a global ecosystem composed of living organisms (biota) and the abiotic (nonliving) factors from which they derive energy and nutrients.

Before the coming of life, Earth was a bleak place, a rocky globe with shallow seas and a thin band of gases—largely carbon dioxide, carbon monoxide, molecular nitrogen, hydrogen sulfide, and water vapour. It was a hostile and barren planet. This strictly inorganic state of the Earth is called the geosphere; it consists of the lithosphere (the rock and soil), the hydrosphere (the water), and the atmosphere (the air). Energy from the Sun relentlessly bombarded the surface of the primitive Earth, and in time—millions of years—chemical and physical actions produced the first evidence of life: formless, jellylike blobs that could collect energy from the environment and produce more of their own kind. This generation of life in the thin outer layer of the geosphere established what is called the biosphere, the "zone of life," an energy-diverting skin that uses the matter of the Earth to make living substance.

The biosphere is a system characterized by the continuous cycling of matter and an accompanying flow of solar energy in which certain large molecules and cells are self-reproducing. Water is a major predisposing factor, for all life depends on it. The elements carbon, hydrogen, nitrogen, oxygen, phosphorus, and sulfur, when combined as proteins, lipids, carbohydrates, and nucleic acids, provide the building blocks, the fuel, and the direction for the creation of life. Energy flow is required to maintain the structure of organisms by the formation and splitting of phosphate bonds. Organisms are cellular in nature and always contain some sort of enclosing membrane structure, and all have nucleic acids that store and transmit genetic information.

All life on Earth depends ultimately upon green plants, as well as upon water. Plants utilize sunlight in a process called photosynthesis to produce the food upon which animals feed and to provide, as a by-product, oxygen, which most animals require for respiration. At first, the oceans and the lands were teeming with large numbers of a few kinds of simple single-celled organisms, but slowly plants and animals of increasing complexity evolved. Interrelationships developed so that certain plants grew in association with certain other plants, and animals associated with the plants and with one another to form communities of organisms, including those of forests, grasslands, deserts, dunes, bogs, rivers, and lakes. Living communities and their nonliving environment are inseparably interrelated and constantly interact upon each other. For convenience, any segment of the landscape that includes the biotic and abiotic components is called an ecosystem. A lake is an ecosystem when it is considered in totality as not just water but also nutrients, climate, and all of the life contained within it. A given forest, meadow, or river is likewise an ecosystem. One ecosystem grades into another along zones termed ecotones, where a mixture of plant and animal species from the two ecosystems occurs. A forest considered as an ecosystem is not simply a stand of trees but is a complex of soil, air, and water, of climate and minerals, of bacteria, viruses, fungi, grasses, herbs, and trees, of insects, reptiles, amphibians, birds, and mammals.

Stated another way, the abiotic, or nonliving, portion of each ecosystem in the biosphere includes the flow of energy, nutrients, water, and gases and the concentrations of organic and inorganic substances in the environment. The biotic, or living, portion includes three general categories of organisms based on their methods of acquiring energy: the primary producers, largely green plants; the consumers, which include all the animals; and the decomposers, which include the microorganisms that break down the remains of plants and animals into simpler components for recycling in the biosphere. Aquatic ecosystems are those involving marine environments and freshwater environments on the land. Terrestrial ecosystems are those based on major vegetational types, such as forest, grassland, desert, and tundra. Particular kinds of animals are associated with each such plant province.

Ecosystems may be further subdivided into smaller biotic units called communities. Examples of communities include the organisms in a stand of pine trees, on a coral reef, and in a cave, a valley, a lake, or a stream. The major consideration in the community is the living component, the organisms; the abiotic factors of the environment are excluded.

A community is a collection of species populations. In a stand of pines, there may be many species of insects, of birds, of mammals, each a separate breeding unit but each dependent on the others for its continued existence. A species, furthermore, is composed of individuals, single functioning units identifiable as organisms. Beyond this level, the units of the biosphere are those of the organism: organ systems composed of organs, organs of tissues, tissues of cells, cells of molecules, and molecules of atomic elements and energy. The progression, therefore, proceeding upward from atoms and energy, is toward fewer units, larger and more complex in pattern, at each successive level.

Resources of the Biosphere

The Flow of Energy

The photosynthetic process Life on earth depends on the harnessing of solar energy by the process of photosynthesis. Photosynthetic plants convert solar energy into the chemical energy of living tissue, and that stored chemical energy flows into herbivores, predators, parasites, decomposers, and all other forms of life. In the photosynthetic process, light energy is absorbed by the chlorophyll molecules of plants to convert carbon dioxide and water into carbohydrates and oxygen gas. Proteins, fats, nucleic acids, and other compounds also are synthesized during the process, as long as elements such as nitrogen, sulfur, and phosphorus are available.

Efficiency of Solar Energy Utilization

Most solar energy occurs at wavelengths unsuitable for photosynthesis. Between 98 and 99 percent of solar energy reaching the Earth is reflected from leaves and other surfaces and absorbed by other molecules, which convert it to heat. Thus, only 1 to 2 percent is available to be captured by plants. The rate at which plants photosynthesize depends on the amount of light reaching the leaves, the temperature of the environment, and the availability of water and other nutrients such as nitrogen and phosphorus. The measurement of the rate at which organisms convert light energy (or inorganic chemical energy) to the chemical energy of organic compounds is called primary productivity. Hence, the total amount of energy assimilated by plants in an ecosystem during photosynthesis (gross primary productivity) varies among environments. (Productivity is often measured by an increase in biomass, a term used to refer to the weight of all living organisms in an area. Biomass is reported in grams or metric tons.)

Much of the energy assimilated by plants through photosynthesis is not stored as organic material but instead is used during cellular respiration. In this process organic compounds such as carbohydrates, proteins, and fats are broken down, or oxidized, to provide energy (in the form of adenosine triphosphate [ATP]) for the cell's metabolic needs. The energy not used in this process is stored in plant tissues for further use and is called net primary productivity. About 40 to 85 percent of gross primary productivity is not used during respiration and becomes net primary productivity. The highest net primary productivity in terrestrial environments occurs in swamps and marshes and tropical rainforests; the lowest occurs in deserts. In aquatic environments, the highest net productivity occurs in estuaries, algal beds, and reefs. Consequently, these environments are especially critical for the maintenance of worldwide biological productivity.

Energy Transfers and Pyramids

A small amount of the energy stored in plants, between 5 and 25 percent, passes into herbivores (plant eaters) as they feed, and a similarly small percentage of the energy in herbivores then passes into carnivores (animal eaters). The result is a pyramid of energy, with most energy concentrated in the photosynthetic organisms at the bottom of food chains and less energy at each higher trophic level (a division based on the main nutritional source of the organism;). Some of the remaining energy does not pass directly into the plant-herbivore-carnivore food chain but instead is diverted into the detritus food chain. Bacteria, fungi, scavengers, and carrion eaters that consume detritus (detritivores) are all eventually consumed by other organisms.

The rate at which these consumers convert the chemical energy of their food into their own biomass is called secondary productivity. The efficiency at which energy is transferred from one trophic level to another is called ecological efficiency. On average it is estimated that there is only a 10 percent transfer of energy.

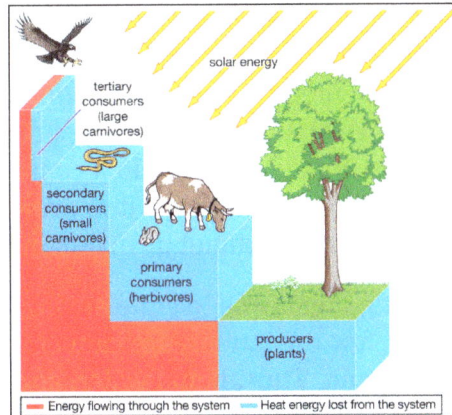

Figure: Transfer of energy through an ecosystem. At each trophic level only a
small proportion of energy (approximately 10 percent) is transferred to the next level.

Energy is lost in several ways as it flows along these pathways of consumption. Most plant tissue is uneaten by herbivores, and this stored energy is therefore lost to the plant-herbivore-carnivore food chain. In terrestrial communities less than 10 percent of plant tissue is actually consumed by herbivores. The rest falls into the detritus pathway, although the detritivores consume only some of this decaying tissue. Oil and coal deposits are major repositories of this unused plant energy and have accumulated over long periods of geologic time.

The efficiency by which animals convert the food they ingest into energy for growth and reproduction is called assimilation efficiency. Herbivores assimilate between 15 and 80 percent of the plant material they ingest, depending on their physiology and the part of the plant that they eat. For example, herbivores that eat seeds and young vegetation high in energy have the highest assimilation efficiencies, those that eat older leaves have intermediate efficiencies, and those that feed on decaying wood have very low efficiencies. Carnivores generally have higher assimilation efficiencies than herbivores, often between 60 and 90 percent, because their food is more easily digested.

The overall productivity of the biosphere is therefore limited by the rate at which plants convert solar energy (about 1 percent) into chemical energy and the subsequent efficiencies at which other organisms at higher trophic levels convert that stored energy into their own biomass (approximately 10 percent). Human-induced changes in net primary productivity in the parts of the biosphere that have the highest productivity, such as estuaries and tropical moist forests, are likely to have large effects on the overall biological productivity of the Earth.

Environmental Conditions

Most organisms are limited to either a terrestrial or an aquatic environment. An organism's ability to tolerate local conditions within its environment further restricts its distribution. One parameter, such as temperature tolerance, may be important in determining the limits of distribution, but often a combination of variables, such as temperature tolerance and water requirements, is important. Extreme environmental variables can evoke physiological and behavioral responses from organisms. The physiological response helps the organism maintain a constant internal environment (homeostasis), while a behavioral response allows it to avoid the environmental challenge—a fallback strategy if homeostasis cannot be maintained.

The ways in which modern living organisms tolerate environmental conditions reflect the aquatic origins of life. With few exceptions, life cannot exist outside the temperature range at which water is a liquid. Thus, liquid water, and temperatures that maintain water as a liquid, are essential for sustaining life. Within those parameters, the concentrations of dissolved salts and other ions, the abundance of respiratory gases, atmospheric or hydrostatic pressure, and rate of water flow all influence the physiology, behaviour, and distribution of organisms.

Temperature

Temperature has the single most important influence on the distribution of organisms because it determines the physical state of water. Most organisms cannot live in conditions in which the temperature remains below 0 °C or above 45 °C for any length of time. Adaptations have enabled certain species to survive outside this range—thermophilic bacteria have been found in hot springs in which the temperatures may approach the boiling point, and certain polar mosses and lichens can tolerate temperatures of −70 °C—but these species are the exceptions. Few organisms can remain for long periods at temperatures above 45 °C, because organic molecules such as proteins will begin to denature. Nor are temperatures below freezing conducive to life: cells will rupture if the water they contain freezes.

Most organisms are not able to maintain a body temperature that is significantly different from that of the environment. Sessile organisms, such as plants and fungi, and very small organisms and animals that cannot move great distances, therefore, must be able to withstand the full range of temperatures sustained by their habitat. In contrast, many mobile animals employ behavioral mechanisms to avoid extreme conditions in the short term. Such behaviours vary from simply moving short distances out of the Sun or an icy wind to large-scale migrations.

Some types of animals employ physiological mechanisms to maintain a constant body temperature, and two categories are commonly distinguished: the term cold-blooded is understood to refer to reptiles and invertebrates, and warm-blooded is generally applied to mammals and birds. These terms, however, are imprecise; the more accurate terms, ectotherm for cold-blooded and endotherm for warm-blooded, are more useful in describing the thermal capabilities of these animals. Ectotherms rely on external sources of heat to regulate their body temperatures, and endotherms thermoregulate by generating heat internally.

Terrestrial ectotherms utilize the complex temperature profile of the terrestrial environment to derive warmth. They can absorb solar radiation, thus raising their body temperatures above that of the surrounding air and substrate, unlike the aquatic ectotherm, whose body temperature is usually very close to that of the environment. As this solar radiation is taken up, physiological mechanisms contribute to the regulation of heat—peripheral blood vessels dilate and heart rate increases. The animal also may employ behavioral mechanisms, such as reorienting itself toward the Sun or flattening its body and spreading its legs to maximize its surface area exposure. At night, loss of heat may be reduced by other behavioral and physiological mechanisms—the heart rate may slow, peripheral blood vessels may constrict, surface area may be minimized, and shelter may be sought.

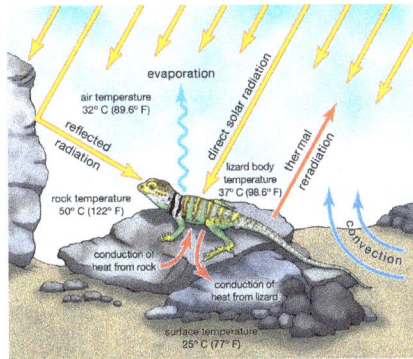

Figure: Energy exchange between a terrestrial reptile and the environment.

Endotherms maintain body temperature independently of the environment by the metabolic production of heat. They generate heat internally and control passive heat loss by varying the quality of their insulation or by repositioning themselves to alter their effective surface area (i.e., curling into a tight ball). If heat loss exceeds heat generation, metabolism increases to make up the loss. If heat generation exceeds the rate of loss, mechanisms to increase heat loss by evaporation occur. In either case, behavioral mechanisms can be employed to seek a more suitable thermal environment.

To survive for a limited period in adverse conditions, endotherms may employ a combination of behavioral and physiological mechanisms. In cold weather, which requires an increase in energy consumption, the animal may enter a state of torpor in which its body temperature, metabolism, respiratory rate, and heart rate are depressed. Long-term winter hypothermia, or hibernation, is an extended state of torpor that some animals use as a response to cold conditions. Torpor and hibernation free the animals from energetically expensive maintenance of high body temperatures, saving energy when food is limited.

Another form of torpor, estivation, is experienced by animals in response to heat stress. This state is seen more often in ectothermic animals than in endotherms, but in both the stimulus for estivation is usually a combination of high temperatures and water shortage.

Humidity

Most terrestrial organisms must maintain their water content within fairly narrow limits. Water commonly is lost to the air through evaporation or, in plants, transpiration. Because most water loss occurs by diffusion and the rate of diffusion is determined by the gradient across the diffusion barrier such as the surface of a leaf or skin, the rate of water loss will depend on the relative humidity of the air. Relative humidity is the percent saturation of air relative to its total saturation possible at a given temperature. When air is totally saturated, relative humidity is said to be 100 percent. Cool air that is completely saturated contains less water vapour than completely saturated warm air because the water vapour capacity of warm air is greater. Diffusion gradients across skin or leaves, therefore, can be much steeper in summer when the air is warm, rendering evaporative water loss a much more serious problem in warm environments than in cool environments. Nevertheless, rates of water loss are higher in dry air (conditions of low relative humidity) than in moist air (conditions of high relative humidity), regardless of the temperature.

Water loss from evaporation must be compensated by water uptake from the environment. For most plants, transpirational water loss is countered by the uptake of water from the soil via roots. For animals, water content can be replenished by eating or drinking or by uptake through the integument. For organisms living in dry environments, there are many morphological and physiological mechanisms that reduce water loss. Desert plants, or xerophytes, typically have reduced leaf surface areas because leaves are the major sites of transpiration. Some xerophytes shed their leaves altogether in summer, and some are dormant during the dry season.

Desert animals typically have skin that is relatively impervious to water. The major site of evaporation is the respiratory exchange surface, which must be moist to allow the gaseous exchange of oxygen and carbon dioxide. A reduction in amount of water lost through respiration can occur if the temperature of the exhaled air is lower than the temperature of the body. As many animals, such as gazelles, inhale warm air, heat and water vapour from the nasal passages evaporate, cooling the nose and the blood within it. The cool venous blood passes close to and cools the warm arterial blood traveling to the brain. If the brain does not require cooling, the venous blood returns to the heart by another route. The nasal passages also cool the warm, saturated air from the lungs so that water condenses in the nose and is reabsorbed rather than lost to the environment.

pH

The relative acidity or alkalinity of a solution is reported by the pH scale, which is a measure of the concentration of hydrogen ions in solution. Neutral solutions have a pH of 7. A pH of less than 7 denotes acidity (an increased hydrogen ion concentration), and above 7 alkalinity (a decreased hydrogen ion concentration). Many important molecular processes within the cells of organisms occur within a very narrow range of pH. Thus, maintenance of internal pH by homeostatic mechanisms is vital for cells to function properly. Although pH may differ locally within an organism, most tissues are within one pH unit of neutral. Because aquatic organisms generally have somewhat permeable skins or respiratory exchange surfaces, external conditions can influence internal pH. These organisms may accomplish the extremely important task of regulating internal pH by exchanging hydrogen ions for other ions, such as sodium or bicarbonate, with the environment.

The pH of naturally occurring waters can range from very acidic conditions of about 3 in peat swamps to very alkaline conditions of about 9 in alkaline lakes. Naturally acidic water may result from the presence of organic acids, as is the case in a peat swamp, or from geologic conditions such as sulfur deposits associated with volcanic activity. Naturally occurring alkaline waters usually result from inorganic sources. Most organisms are unable to live in conditions of extreme alkalinity or acidity.

Salinity

The term salinity refers to the amount of dissolved salts that are present in water. Sodium and chloride are the predominant ions in seawater, and the concentrations of magnesium, calcium, and sulfate ions are also substantial. Naturally occurring waters vary in salinity from the almost pure water, devoid of salts, in snowmelt to the saturated solutions in salt lakes such as the Dead Sea. Salinity in the oceans is constant but is more variable along the coast where seawater is diluted with freshwater from runoff or from the emptying of rivers. This brackish water forms a barrier separating marine and freshwater organisms.

The cells of organisms also contain solutions of dissolved ions, but the range of salinity that occurs in tissues is more narrow than the range that occurs in nature. Although a minimum number of ions must be present in the cytoplasm for the cell to function properly, excessive concentrations of ions will impair cellular functioning. Organisms that live in aquatic environments and whose integument is permeable to water, therefore, must be able to contend with osmotic pressure. This pressure arises if two solutions of unequal solute concentration exist on either side of a semipermeable membrane such as the skin. Water from the solution with a lower solute concentration will cross the membrane diluting the more highly concentrated solution until both concentrations are equalized. If the salt concentration of an animal's body fluids is higher than that of the surrounding environment, the osmotic pressure will cause water to diffuse through the skin until the concentrations are equal unless some mechanism prevents this from happening.

Many marine invertebrates have the same osmotic pressure as seawater. When the salt concentration of their surroundings changes, however, they must be able to adjust. Two means of contending with this situation are employed, and, depending on how they regulate the salt concentrations of their tissues, organisms are classified as osmoregulators or osmoconformers. The osmotic concentration of the body fluids of an osmoconformer changes to match that of its external environment, whereas an osmo-regulator controls the osmotic concentration of its body fluids, keeping them constant in spite of external alterations. Aquatic organisms that can tolerate a wide range of external ion concentrations are called euryhaline; those that have a limited tolerance are called stenohaline.

Even if aquatic organisms have an integument that is relatively impermeable to water, as well as to small inorganic ions, their respiratory exchange surfaces are permeable. Hence, organisms occurring in water that has a lower solute concentration than their tissues (e.g., trout in mountain streams) will constantly lose ions to the environment as water flows into their tissues. In contrast, organisms in salty environments face a constant loss of water and an influx of ions.

Many mechanisms have evolved that deal with these problems. Because water cannot be readily pumped across cell membranes, salinity balance is usually maintained by actively transporting inorganic ions, usually sodium and chloride. This process con-sumes energy and can usurp a large portion of the energy budget of animals in very saline environments. In marine fish, gill cells pump ions out of the body into the sea, while in freshwater fish gill cells pump ions in the opposite direction. Passive water loss in marine fish is compensated primarily in one of two ways. Most bony fish drink copi-ously and excrete salt across the gills, while the majority of sharks artificially elevate the salt concentration of their tissues above that of seawater with urea and other organic molecules, allowing water to slowly and passively enter the body. Through their food and across their gills, freshwater fish replenish most of the ions they lose. They also produce large quantities of very dilute urine to excrete excess water that diffuses into their bodies.

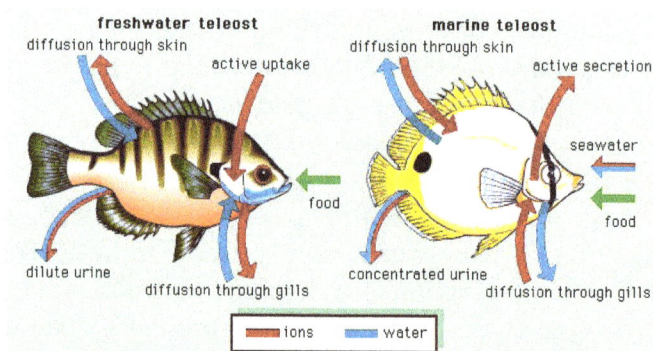

Figure: Osmotic regulation in freshwater and marine teleost fish

Water Currents

The flow of water presents special problems for aquatic organisms. Flow is associated with rivers, oceanic currents, and waves and can be laminar (streamlined) or turbulent. Many organisms are specialized to live in flowing environments; the main obstacle to this lifestyle is the constant threat of being washed away. Both plants and animals have evolved mechanisms that help to anchor them to the substratum in flowing water (e.g., the holdfast of kelp or the byssus threads of mussels). If anchorage can be assured, there are many advantages to living in this environment. Flowing water generally is well oxygenated, and the supply is continuous; nutrients and food are constantly replenished as well. The very precariousness of the environment also affords some protection from predation because the number of predators that make this type of habitat home is limited.

Pressure

Atmospheric Pressure

Variations in atmospheric pressure can present special problems for the respiratory systems of animals because atmospheric pressure affects the exchange of oxygen and carbon dioxide that occurs during animal respiration. Normal atmospheric pressure at sea level is the total pressure that a column of air above the surface of the Earth exerts (760 millimetres of mercury, or 1 atmosphere). The total pressure is the sum of the pressures that each gas—mainly nitrogen, oxygen, and carbon dioxide—would exert alone (the partial pressure of that gas; see respiration: The gases in the environment). As an animal breathes, oxygen moves from the environment across the respiratory surfaces into the blood; carbon dioxide moves in the reverse direction. This process occurs primarily by passive diffusion; each gas moves from an area of greater to lesser partial pressure, driven by the differential that exists across the respiratory surface. At higher altitudes, where the atmospheric pressure is lower, the partial pressure of oxygen is also lower. The partial pressure differential of oxygen, therefore, is also lower, and the organism effectively receives less oxygen when it breathes, even though the percentage of oxygen in the air remains constant. This lack of oxygen is why humans carry oxygen when ascending to high altitudes. Humans who live in mountainous regions, however, can become acclimatized to the lowered availability of oxygen, and certain animals such as llamas have adaptations of the blood that allow them to live at high altitudes. Birds have very efficient lungs, and many apparently have no problems flying to high altitudes, even for extended flights.

Hydrostatic Pressure

Because air and water have vastly different densities, the pressures experienced in terrestrial and aquatic habitats differ markedly. A column of water, so much denser than air, exerts a greater amount of pressure than a column of air. With each 10-metre (32.8-foot) increase in depth, there is an increase in hydrostatic pressure equivalent to one atmosphere. Mean ocean depth is about 3,800 metres and has a pressure of about 380 atmospheres. To surmount this environmental challenge, animals that live at great depths lack air compartments such as lungs or swim bladders. Surface-dwelling animals that dive to great depths meet this challenge differently. As pressure increases during a dive, air compartments compress, returning to their former volume when the animal surfaces. Air is forced into the trachea, bronchi, and bronchioles, where no gas uptake occurs. Thus, the increased pressure cannot drive more gases into the bloodstream, and, as the animal rises, it does not experience the "bends" (decompression sickness resulting from a rapid reduction of air pressure). In contrast, sea snakes avoid the bends by excreting nitrogen across the skin to offset the uptake of this gas from the lungs.

Hydrosphere

The hydrosphere is the component of the Earth that is composed of all liquid water found on the planet. The hydrosphere includes water storage areas such as oceans, seas, lakes, ponds, rivers, and streams. Overall, the hydrosphere is very large, with the oceans alone covering about 71% of the surface area of Earth.

The motion of the hydrosphere and the exchange of water between the hydrosphere and cryosphere is the basis of the hydrologic cycle. The continuous movement and exchange of water helps to form currents that move warm water from the tropics to the poles and help regulate the temperature of the Earth. The exchanging of water is thus a vital part of the hydrosphere.

It is important to note that although the hydrosphere is primarily composed of water, there are also some "impurities" or additions to this water that include dissolved minerals, dissolved gases, and particulates. Some of these can be considered pollution, while others are necessary for health of ecosystems. For example, too much sediment is harmful to the surrounding ecosystems, while insufficient levels of dissolved oxygen in the water lead to hypoxic conditions that can harm ecosystems. Thus a delicate balance is needed for healthy ecosystems that surround different components of the hydrosphere.

Figure: The spring shown above would be considered part of the hydrosphere.

Components

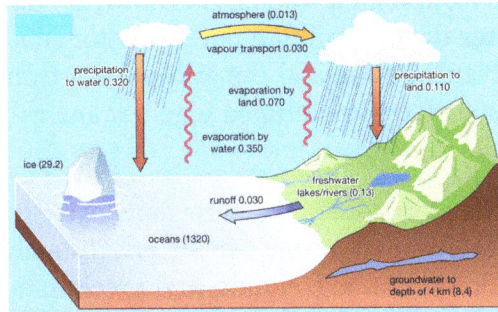

Figure: This image summarizes the hydrosphere cycle

Any water storage area on the Earth that holds liquid water is considered to be a part of the hydrosphere. Because of this, there is an extensive list of formations that make up the hydrosphere. These include:

Oceans: Most of the water on the planet Earth is salt water, and the vast majority of this salt water is held in the oceans.

- Fresh water: Fresh water is much less abundant than salt water, and is held in a variety of different places.

  ◦ Surface water: Surface sources of freshwater include lakes, rivers, and streams.

  ◦ Ground water: Fresh water held beneath ground makes up a small portion of the fresh water on Earth.

- Glacial water: Water that melts off of glaciers.

- Atmospheric water vapour.

Water Cycle

Water cycle or hydrologic cycle refers to the cycle that involves the continuous circulation of water in the Earth-atmosphere system. Of the many processes involved in the water cycle, the most important are evaporation, transpiration, condensation, precipitation, and runoff. Although the total amount of water within the cycle remains essentially constant, its distribution among the various processes is continually changing.

Evaporation, one of the major processes in the cycle, is the transfer of water from the surface of the Earth to the atmosphere. By evaporation, water in the liquid state is transferred to the gaseous, or vapour, state. This transfer occurs when some molecules in water mass have attained sufficient kinetic energy to eject themselves from the water surface. The main factors affecting evaporation are temperature, humidity, wind speed, and solar radiation. The direct measurement of evaporation, though desirable, is difficult and possible only at point locations. The principal source of water vapour is the oceans, but evaporation also occurs in soils, snow, and ice. Evaporation from snow and ice, the direct conversion from solid to vapour, is known as sublimation. Transpiration is the evaporation of water through minute pores, or stomata, in the leaves of plants. For practical purposes, transpiration and the evaporation from all water, soils, snow, ice, vegetation, and other surfaces are lumped together and called evapotranspiration, or total evaporation.

Water vapour is the primary form of atmospheric moisture. Although its storage in the atmosphere is comparatively small, water vapour is extremely important in forming the moisture supply for dew, frost, fog, clouds, and precipitation. Practically all water vapour in the atmosphere is confined to the troposphere (the region below 6 to 8 miles (10 to 13 km) altitude).

The transition process from the vapour state to the liquid state is called condensation. Condensation may take place as soon as the air contains more water vapour than it can receive from a free water surface through evaporation at the prevailing temperature. This condition occurs as the consequence of either cooling or the mixing of air masses of different temperatures. By condensation, water vapour in the atmosphere is released to form precipitation.

Precipitation that falls to the Earth is distributed in four main ways: some is returned to the atmosphere by evaporation, some may be intercepted by vegetation and then evaporated from the surface of leaves, some percolates into the soil by infiltration, and the remainder flows directly as surface runoff into the sea. Some of the infiltrated precipitation may later percolate into streams as groundwater runoff. Direct measurement of runoff is made by stream gauges and plotted against time on hydrographs.

Most groundwater is derived from precipitation that has percolated through the soil. Groundwater flow rates, compared with those of surface water, are very slow and variable, ranging from a few millimetres to a few metres a day. Groundwater movement is studied by tracer techniques and remote sensing.

Ice also plays a role in the water cycle. Ice and snow on the Earth's surface occur in various forms such as frost, sea ice, and glacier ice. When soil moisture freezes, ice also occurs beneath the Earth's surface, forming permafrost in tundra climates. About 18,000 years ago glaciers and ice caps covered approximately one-third of the Earth's land surface. Today about 12 percent of the land surface remains covered by ice masses.

Human Impacts on Hydrosphere

In recent history humans have drastically changed the hydrosphere. Water pollution, river damming,

Wetland drainage, climate change, and irrigation have all changed the hydrosphere. Eutrophication caused by the release of fertilizers and sewage into water storage areas has caused aquatic environments to be artificially enriched with nutrients. The excessive algal blooms can result in harmful hypoxic conditions in the water. Acid rain from SO_x and NO_x emissions from fossil fuel combustion has resulted in the acidification of components of the hydrosphere, harming surrounding ecosystems.

Finally, when humans change the natural flow of water in the hydrosphere by diverting and damming rivers it harms surrounding ecosystems that rely on the water source. This can also result in the drying out of some aquatic areas and excessive amounts of sediment entering streams and rivers.

Geosphere

The geosphere includes the rocks and minerals on Earth – from the molten rock and heavy metals in the deep interior of the planet to the sand on beaches and peaks of mountains. The geosphere also includes the abiotic (non-living) parts of soils, and the skeletons of animals that may become fossilized over geologic time.

Beyond these parts, the geosphere is about processes. The processes of the rock cycle such as metamorphism, melting and solidification, weathering, erosion, deposition, and burial are responsible for a constant recycling of rocks on Earth between sedimentary, igneous, and metamorphic states.

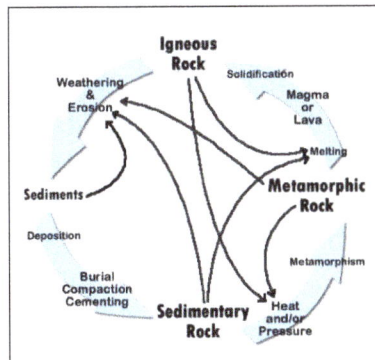

The rock cycle.

- Sedimentary rocks are formed via weathering and transport of existing rocks, and then deposition, cementation, and compaction into a sedimentary rock.

- Igneous rocks are formed by cooling and crystallization of molten rock.

- Metamorphic rocks are formed when heat or pressure are applied to other rocks.

The primary agent driving these processes is the movement of Earth's tectonic plates, which creates mountains, volcanoes, and ocean basins. Changes in the rate that rocks are made and destroyed can have a profound effect on the planet. As the rate of plate tectonic movements has changed over geologic time scales, the rock cycle has changed as well, and these changes have been able to affect climate. For example, at times when the rate of plate movements has been high, there is more volcanic activity, which releases more particles into the atmosphere. Faster plate tectonic movements also mean more mountains are built in areas where plates converge. As rocks are uplifted into mountains, they start to erode and dissolve, sending sediments and nutrients into waterways and impacting the ecosystems for living things.

As climate changes the geosphere interacts with various other parts of the Earth system:

Biosphere: The carbon cycle, usually linked with the Earth's biosphere, includes deep storage of carbon in the form of fossil fuels like coal, oil, and gas as well as carbonate rocks like limestone. The carbon cycle is one of several biogeochemical cycles, which all involve the geosphere, the biosphere, and other spheres of the Earth system.

- Cryosphere: Glaciers and ice sheets, parts of the cryosphere, have a large impact on the rocks and sediments below them. For example, the continental ice sheet moved rocks as it flowed south during the last ice age, creating Cape Cod, Long Island, hills, and lakes. The ice is also able to have a regional affect on the elevation of land, which lifts up once ice has melted from its surface. The land in north central Canada has been slowly lifting up after the melt of glaciers from the last ice age.

- Hydrosphere and Atmosphere: The erosion of rocks, a major part of the rock cycle and change in the geosphere over time, turns rock to sediment and then, sometimes, to sedimentary rock. But erosion, transportation, and deposition of sediments wouldn't occur without the hydrosphere's rivers, lakes, and ocean or the atmosphere's winds and precipitation. Different combinations of sedimentary rocks form in environments with different climate conditions. This allows geologists to reconstruct what an environment was like millions of years ago based on the sedimentary rocks that were deposited.

References

- Earth's-Spheres-and-atmosphere: cotf.edu, Retrieved 11 August, 2019

- Atmosphere, science: britannica.com, Retrieved 2 February, 2019

- Biosphere, science: britannica.com, Retrieved 29 June, 2019

- Hydrosphere: energyeducation.ca, Retrieved 30 March, 2019

- Water-cycle, science: britannica.com, Retrieved 23 July, 2019

- Hydrosphere: energyeducation.ca, Retrieved 16 January, 2019

4

Earth's Cycles

There are numerous cycles which are involved in regulating and balancing the Earth and its atmosphere. A few of them are oxygen cycle, carbon cycle, nitrogen cycle, sulfur cycle and phosphorus cycle. This chapter has been carefully written to provide an easy understanding of the varied facets of these Earth's cycles.

The Earth is a dynamic planet. Geological and biological processes cause energy and the elements necessary for life-carbon, hydrogen, nitrogen, oxygen, sulfur, and phosphorus- to circulate through global "reservoirs." These reservoirs are the biosphere (the living portion of our planet), the atmosphere, the ocean, and the solid Earth. It is only because of this cycling that life can thrive. The cycling of elements determines the environment, for example by regulating the composition, and thus the temperature, of the atmosphere.

Oxygen Cycle

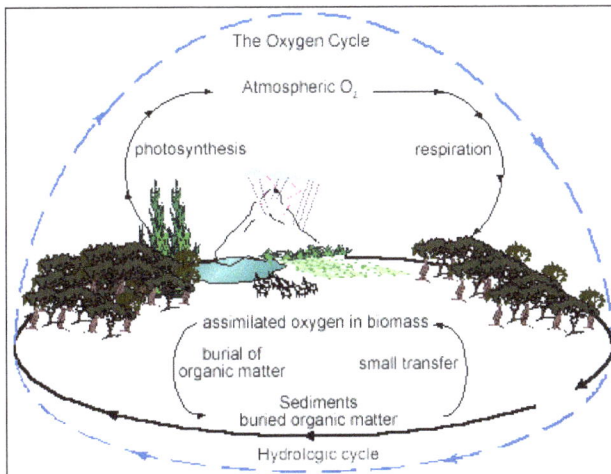

The oxygen cycle is the cycle that helps move oxygen through the three main regions of the Earth, the Atmosphere, the Biosphere, and the Lithosphere. The Atmosphere is of course the region of gases that lies above the Earth's surface and it is one of the largest reservoirs of free oxygen on earth. The Biosphere is the sum of all the Earth's ecosystems. This also has some free oxygen produced from photosynthesis and other life processes. The largest reservoir of oxygen is the lithosphere. Most of this oxygen is not on its own or free moving but part of chemical compounds such as silicates and oxides.

The atmosphere is actually the smallest source of oxygen on Earth comprising only 0.35% of the Earth's total oxygen. The smallest comes from biospheres. The largest is as mentioned before in the Earth's crust. The Oxygen cycle is how oxygen is fixed for freed in each of these major regions.

In the atmosphere Oxygen is freed by the process called photolysis. This is when high energy sunlight breaks apart oxygen bearing molecules to produce free oxygen. One of the most wellknown photolysis it the ozone cycle. O_2 oxygen molecule is broken down to atomic oxygen by the ultra violet radiation of sunlight. This free oxygen then recombines with existing O_2 molecules to make O_3 or ozone. This cycle is important because it helps to shield the Earth from the majority of harmful ultra violet radiation turning it to harmless heat before it reaches the Earth's surface.

In the biosphere the main cycles are respiration and photosynthesis. Respiration is when animals and humans breathe consuming oxygen to be used in metabolic process and exhaling carbon dioxide. Photosynthesis is the reverse of this process and is mainly done by plants and plankton.

The lithosphere mostly fixes oxygen in minerals such as silicates and oxides. Most of the time the process is automatic all it takes is a pure form of an element coming in contact with oxygen such as what happens when iron rusts. A portion of oxygen is freed by chemical weathering. When oxygen bearing mineral is exposed to the elements a chemical reaction occurs that wears it down and in the process produces free oxygen.

These are the main oxygen cycles and each play an important role in helping to protect and maintain life on the Earth.

Carbon Cycle

Carbon is the fourth most abundant element in the Universe, after hydrogen, helium, and oxygen. Carbon is a fundamental building block of life; life on Earth is comprised of carbon-based life forms. Carbon also cycles through the oceans and the biosphere over both short and long-term time scales. The geological carbon cycle takes place over hundreds of millions of years and involves the cycling of carbon through the various layers of the Earth. The biological/physical carbon cycle occurs over days, weeks, months, and years and involves the absorption, conversion, and respiration of carbon by living organisms.

Biological Carbon Cycle

Photosynthesis traps carbon dioxide from the atmosphere to produce glucose and it stores energy. Glucose, of course, is used to make other organic molecules and is used as a source of energy in respiration.

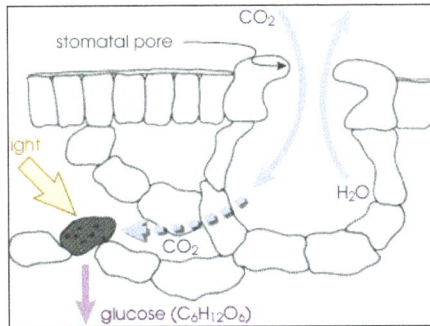

In respiration and in the oxidative decomposition of plant materials, the carbon in organic molecules is converted to CO_2. Only a very small percentage of the organic carbon is sequestered in sediments.

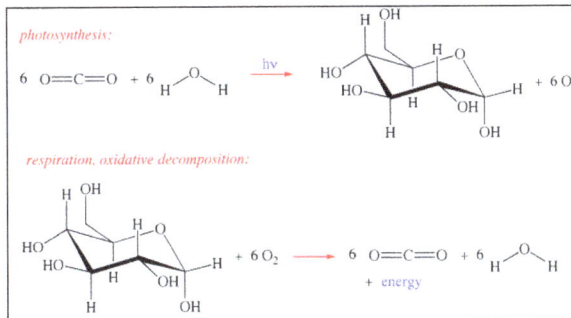

The biological carbon cycle is not only faster than the geological carbon cycle. The amount of carbon taken up by photosynthesis and released back to the atmosphere by respiration each year is 1,000 times greater than the amount of carbon that moves through the geological cycle on an annual basis.

The biological carbon cycle plays a role in the long-term, geological cycling of carbon. The presence of land vegetation enhances the weathering of soil, leading to the uptake of carbon dioxide from the atmosphere. In the oceans, some of the carbon taken up by phytoplankton is used to make shells of calcium carbonate that settle to the bottom after the organisms die to form sediments. Marine animals, such as corals, also use dissolved carbon dioxide in biomineralization.

During the daytime in the growing season, leaves absorb sunlight and take up carbon dioxide from the atmosphere. Plants, animals and soil microbes consume the carbon in organic matter and return carbon dioxide to the atmosphere.

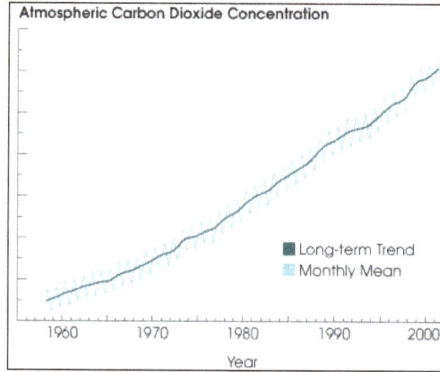

When conditions are too cold or too dry, photosynthesis and respiration cease along with the movement of carbon between the atmosphere and the land surface. The amounts of carbon that move from the atmosphere through photosynthesis, respiration, and back to the atmosphere are large and produce oscillations in atmospheric carbon dioxide concentrations.

Significant amounts of carbon are stored in the biomass of forests and in the soil. Terrestrial sources release the stored carbon when forests are cleared for agriculture. Organisms in the ocean consume and release large quantities of carbon dioxide but ocean biological carbon cycles are faster than terrestrial cycles. There is virtually no storage of carbon as biomass. Photosynthetic plankton are consumed by zooplankton within days to weeks.

Carbon dioxide exchange in the oceans is controlled by sea surface temperatures, circulating currents, and by the biological processes of photosynthesis and respiration. Carbon dioxide solvation is temperature dependent. Cold ocean temperatures favor the uptake of carbon dioxide from the atmosphere while warm temperatures can cause the ocean surface to release carbon dioxide. Cold, downward moving currents such as those that occur over the North Atlantic absorb carbon dioxide and transfer it to the deep ocean. Upward moving currents such as those in the tropics bring carbon dioxide up from depth and release it to the atmosphere.

The Geological Carbon Cycle

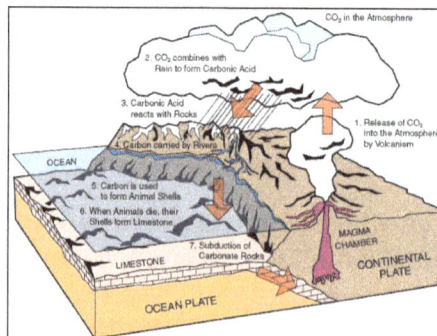

The origin atmosphere of the Earth was rich in reduced gases including methane, CH_4. The carbon content of the Earth steadily increased over eons as a result of collisions with carbon-rich meteors. As the oxygen content of the atmosphere increase, the carbon-containing molecules were oxidized to CO_2.

Carbon dioxide, an acidic oxide, and carbonic acid have slowly but continuously combined with calcium and magnesium oxides, basic oxides, in the crust to form insoluble carbonates.

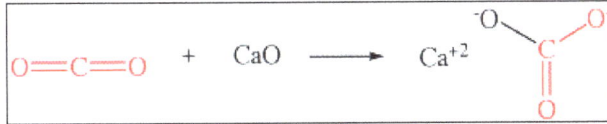

$$O{=}C{=}O \quad + \quad CaO \quad \longrightarrow \quad Ca^{+2} \qquad \text{(calcium carbonate structure)}$$

Carbon dioxide in water, or carbonic acid, also reacts with silicate rock. The chemical weather of calcium silicates by carbonic acid produces calcium carbonate and silicon dioxide.

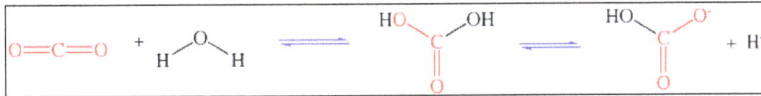

$$O{=}C{=}O \; + \; H_2O \;\rightleftharpoons\; \text{(carbonic acid)} \;\rightleftharpoons\; + \; H^+$$

$$2\,H_2CO_3 \; + \; Ca_2SiO_4 \; \longrightarrow \; SiO_2 \; + \; 2\,CaCO_3 \; + \; 2\,H_2O$$

Through the process of erosion, these carbonates are washed into the ocean and eventually settle to the bottom. These materials are drawn into the mantle by subduction (a process in which one lithospheric plate descends beneath another, often as a result of folding or faulting) at the edges of continental plates. The heat and pressure within the Earth causes the metal carbonates to react with silica to form metal silicates, such as Ca_2SiO_4, and CO_2. The carbon is then returned to the atmosphere as carbon dioxide during volcanic eruptions.

$$2\,CaCO_3 \; + \; SiO_2 \; \longrightarrow \; Ca_2SiO_4 \; + \; 2\,CO_2$$

The balance between weathering, subduction and volcanism controls atmospheric carbon dioxide concentrations over time periods of hundreds of millions of years but the concentration of CO_2 has changed dramatically. The oldest geologic sediments suggest that, before life evolved, the concentration of atmospheric carbon dioxide may have been one-hundred times that of the present, providing a substantial greenhouse effect during a time of low solar output. Ice core samples taken in Antarctica and Greenland have shown that carbon dioxide concentrations during the last ice age were only half of what they are today.

Over the lifetime of the earth, roughly 75 % of the carbon injected into the atmosphere by volcanoes has found its way into deposits of calcium carbonate (limestone). Limestone tends to accumulate on the beds of shallow seas where the acidity of sea water is reduced. The acidity is higher on the deep ocean floor and the shells and skeletons of marine organisms dissolve as fast as they precipitate.

Weathering of limestone deposits by rain tends to return carbon atoms to the short term reservoirs and to atmospheric carbon dioxide. Weathering of silicate rocks by carbonic acid is faster in a warmer climate because rainfall amounts tend to be greater. By providing calcium ions, weathering promotes limestone formation and removal of carbon dioxide from the atmosphere. An increase in average temperature would, eventually, favor decreasing atmospheric carbon dioxide concentrations and reduce global temperatures however this geochemical process is too slow to have an effect on human-produced global warming.

Human Activities

Recently, humans have made some big changes to the Earth's carbon cycle. By burning huge amounts of fossil fuels and cutting down roughly half of the Earth's forests, humans have decreased the Earth's ability to take carbon out of the atmosphere, while releasing large amounts of carbon into the atmosphere that had been stored in solid form as plant matter and fossil fuels.

This means more carbon dioxide in Earth's atmosphere – which is particularly dangerous since carbon dioxide is a "greenhouse gas" that plays a role in regulating the Earth's temperature and weather patterns.

The scientific community has raised alarms that by making significant changes to the Earth's carbon cycle, we may end up changing our climate or other important aspects of the ecosystem we rely upon to survive. As a result, many scientists advocate decreasing the amount of carbon burned by humans by reducing car use and electricity consumption, and advocate for investing in non-burning sources of energy such as solar power and wind power.

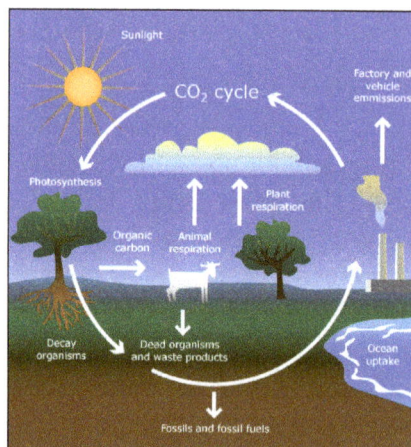

Carbon Cycle.

Carbon Cycle Examples

The carbon cycle consists of many parallel systems which can either absorb or release carbon. Together, these systems work to keep Earth's carbon cycle – and subsequently its climate and biosphere – relatively stable. Below are some examples of parts of Earth's ecosystems that can absorb carbon, turn carbon into living matter, or release carbon back into the atmosphere.

Atmosphere

One major repository of carbon is the carbon dioxide in the Earth's atmosphere. Carbon forms a stable, gaseous molecule in combination with two atoms of oxygen. In nature, this gas is released by volcanic activity, and by the respiration of animals who affix carbon molecules from the food they eat to molecules of oxygen before exhaling it.

Carbon dioxide can be removed from the atmosphere by plants, which take the atmospheric carbon and turn it into sugars, proteins, lipids, and other essential molecules for life. It can also be removed from the atmosphere by absorption into the ocean, whose water molecules can bond with carbon dioxide to form carbonic acid.

Lithosphere

The Earth's crust called the "lithosphere" from the Greek word "litho" for "stone" and "sphere" for globe can also release carbon dioxide into Earth's atmosphere. This gas can be created by chemical reactions in the Earth's crust and mantel.

Volcanic activity can result in natural releases of carbon dioxide. Some scientists believe that widespread volcanic activity may be to blame for the warming of the Earth that caused the Permian extinction.

While the Earth's crust can add carbon to the atmosphere, it can also remove it. Movements of the Earth's crust can bury carbon-containing chemicals such as dead plants and animals deep underground, where their carbon cannot escape back into the atmosphere. Over millions of years, these underground reservoirs of organic matter liquefy and become coal, oil, and gasoline. In recent years, humans have begun releasing much of this sequestered carbon back into the atmosphere by burning these materials to power cars, power plants, and other human equipment.

Biosphere

Among living things, some remove carbon from the atmosphere, while others release it back. The most noticeable participants in this system are plants and animals.

Plants remove carbon from the atmosphere. They don't do this as a charitable act; atmospheric carbon is actually the "food" which plants use to make sugars, proteins, lipids, and other essential molecules for life. Plants use the energy of sunlight, harvested through photosynthesis, to build these organic compounds out of carbon dioxide and other trace elements. Indeed, the term "photosynthesis" comes from the Greek words "photo" for "light" and "synthesis" for "to put together."

In a gracefully balanced set of chemical reactions, animals eat plants (and other animals), and take these synthesized molecules apart again. Animals get their fuel from the chemical energy plants have stored in the bonds between carbon atoms and other atoms during photosynthesis. In order to do that, animal cells dissemble complex molecules such as sugars, fats, and proteins all the way down to single-carbon units – molecules of carbon dioxide, which are produced by reacting carbon-containing food molecules with oxygen from the air.

Oceans

The Earth's oceans have the ability to both absorb and release carbon dioxide. When carbon dioxide from the atmosphere comes into contact with ocean water, it can react with the water molecules to form carbonic acid – a dissolved liquid form of carbon.

When there is more carbonic acid in the ocean compared to carbon dioxide in the atmosphere, some carbonic acid may be released into the atmosphere as carbon dioxide. On the other hand, when there is more carbon dioxide in the atmosphere, more carbon dioxide will be converted to carbonic acid, and ocean acidity levels will rise.

Some scientists have raised concerns that acidity is rising in some parts of the ocean, possibly as a result of increased carbon dioxide in the atmosphere due to human activity. Although these changes in ocean acidity may sound small by human standards, many types of sea life depend on chemical reactions that need a highly specific acidity level to survive. In fact, ocean acidification is currently killing many coral reef communities.

Importance of Carbon Cycle

The carbon cycle, under normal circumstances, works to ensure the stability of variables such as the Earth's atmosphere, the acidity of the ocean, and the availability of carbon for use by living things. Each of its components is of crucial importance to the health of all living things – especially humans, who rely on many food crops and animals to feed our large population.

Carbon dioxide in the atmosphere prevents the sun's heat from escaping into space, very much like the glass walls of a greenhouse. This isn't always a bad thing – some carbon dioxide in the atmosphere is good for keeping the Earth warm and its temperature stable.

But Earth has experienced catastrophic warming cycles in the past, such as the Permian extinction, which is thought to have been caused by a drastic increase in the atmosphere's level of greenhouse gases. No one is sure what caused the change that brought about the Permian extinction. But, greenhouse gases may have been added to an atmosphere by an asteroid impact, volcanic activity, or even massive forest fires.

Whatever the cause, during this warming episode temperatures rose drastically. Much of the Earth became desert and over 90% of all species living at that time went extinct. This is a good example of what can happen if our planet's essential cycles experience a big change.

Nitrogen Cycle

Nitrogen is one of the primary nutrients critical for the survival of all living organisms. It is a necessary component of many biomolecules, including proteins, DNA, and chlorophyll. Although nitrogen is very abundant in the atmosphere as dinitrogen gas (N_2), it is largely inaccessible in this form to most organisms, making nitrogen a scarce resource and often limiting primary productivity in many ecosystems. Only when nitrogen is converted from dinitrogen gas into ammonia (NH_3) does it become available to primary producers, such as plants.

In addition to N_2 and NH_3, nitrogen exists in many different forms, including both inorganic (e.g., ammonia, nitrate) and organic (e.g., amino and nucleic acids) forms. Thus, nitrogen undergoes many different transformations in the ecosystem, changing from one form to another as organisms use it for growth and, in some cases, energy. The major transformations of nitrogen are nitrogen fixation, nitrification, denitrification, anammox, and ammonification. The transformation of nitrogen into its many oxidation states is key to productivity in the biosphere and is highly dependent on the activities of a diverse assemblage of microorganisms, such as bacteria, archaea, and fungi.

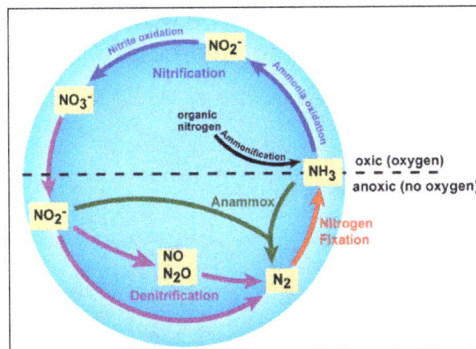

Figure: Major transformations in the nitrogen cycle

Since the mid-1900s, humans have been exerting an ever-increasing impact on the global nitrogen cycle. Human activities, such as making fertilizers and burning fossil fuels, have significantly altered the amount of fixed nitrogen in the Earth's ecosystems. In fact, some predict that by 2030, the amount of nitrogen fixed by human activities will exceed that fixed by microbial processes. Increases in available nitrogen can alter ecosystems by increasing primary productivity and impacting carbon storage. Because of the importance of nitrogen in all ecosystems and the significant impact from human activities, nitrogen and its transformations have received a great deal of attention from ecologists.

Nitrogen Fixation

Nitrogen gas (N_2) makes up nearly 80% of the Earth's atmosphere, yet nitrogen is often the nutrient that limits primary production in many ecosystems. Why is this so? Because plants and animals are not able to use nitrogen gas in that form. For nitrogen to be available to make proteins, DNA, and other biologically important compounds, it must first be converted into a different chemical form. The process of converting N_2 into biologically available nitrogen is called nitrogen fixation. N_2 gas is a very stable compound due to the strength of the triple bond between the nitrogen atoms, and it requires a large amount of energy to break this bond. The whole process requires eight electrons and at least sixteen ATP molecules. As a result, only a select group of prokaryotes are able to carry out this energetically demanding process. Although most nitrogen fixation is carried out by prokaryotes, some nitrogen can be fixed abiotically by lightning or certain industrial processes, including the combustion of fossil fuels.

$$N_2 + 8\,H^+ + 8\,e^- \longrightarrow 2\,NH_3 + H_2$$

Chemical reaction of nitrogen fixation.

Some nitrogen-fixing organisms are free-living while others are symbiotic nitrogen-fixers, which require a close association with a host to carry out the process. Most of the symbiotic associations are very specific and have complex mechanisms that help to maintain the symbiosis. For example, root exudates from legume plants (e.g., peas, clover, soybeans) serve as a signal to certain species of Rhizobium, which are nitrogen-fixing bacteria. This signal attracts the bacteria to the roots, and a very complex series of events then occurs to initiate uptake of the bacteria into the root and trigger the process of nitrogen fixation in nodules that form on the roots.

Nitrogen-fixing nodules on a clover plant root.

Some of these bacteria are aerobic, others are anaerobic; some are phototrophic, others are chemotrophic (i.e., they use chemicals as their energy source instead of light). Although there is great physiological and phylogenetic diversity among the organisms that carry out nitrogen fixation, they all have a similar enzyme complex called nitrogenase that catalyzes the reduction of N2 to NH3 (ammonia), which can be used as a genetic marker to identify the potential for nitrogen fixation. One of the characteristics of nitrogenase is that the enzyme complex is very sensitive to oxygen and is deactivated in its presence. This presents an interesting dilemma for aerobic nitrogen-fixers and particularly for aerobic nitrogen-fixers that are also photosynthetic since they actually produce oxygen. Over time, nitrogen-fixers have evolved different ways to protect their nitrogenase from oxygen. For example, some cyanobacteria have structures called heterocysts that provide a low-oxygen environment for the enzyme and serves as the site where all the nitrogen fixation occurs in these organisms. Other photosynthetic nitrogen-fixers fix nitrogen only at night when their photosystems are dormant and are not producing oxygen.

Genes for nitrogenase are globally distributed and have been found in many aerobic habitats (e.g., oceans, lakes, soils) and also in habitats that may be anaerobic or microaerophilic (e.g., termite guts, sediments, hypersaline lakes, microbial mats, planktonic crustaceans). The broad distribution of nitrogen-fixing genes suggests that nitrogen-fixing organisms display a very broad range of environmental conditions, as might be expected for a process that is critical to the survival of all life on Earth.

Table: Representative prokaryotes known to carry out nitrogen fixation.

Genus	Phylogenetic Affiliation	Lifestyle
Nostoc, Anabaena	Bacteria (Cyanobacteria)	Free-living, aerobic, phototrophic.
Pseudomonas, Azotobacter, Methylomonas	Bacteria	Free-living, aerobic, chemoorganotrophic.
Alcaligenes, Thiobacillus	Bacteria	Free-living, aerobic, chemolithotrophic.
Methanosarcina, Methanococcus	Bacteria	Free-living, anaerobic chemolithotrophic.
Chromatium, Chlorobium	Bacteria	Free-living, anaerobic, phototrophic.
Desulfovibrio, Clostridium	Bacteria	Free-living, anaerobic chemoorganotrophic.
Rhizobium, Frankia	Bacteria	Symbiotic, aerobic, chemoorganotrophic.

Nitrification

Nitrification is the process that converts ammonia to nitrite and then to nitrate and is another important step in the global nitrogen cycle. Most nitrification occurs aerobically and is carried out exclusively by prokaryotes. There are two distinct steps of nitrification that are carried out by distinct types of microorganisms. The first step is the oxidation of ammonia to nitrite, which is carried out by microbes known as ammonia-oxidizers. Aerobic ammonia oxidizers convert ammonia to nitrite via the intermediate hydroxylamine, a process that requires two different enzymes, ammonia monooxygenase and hydroxylamine oxidoreductase. The process generates a very small amount of energy relative to many other types of metabolism; as a result, nitrosofiers are notoriously very slow growers. Additionally, aerobic ammonia oxidizers are also autotrophs, fixing carbon dioxide to produce organic carbon, much like photosynthetic organisms, but using ammonia as the energy source instead of light.

$$NH_3 + O_2 + 2\ e^- \longrightarrow NH_2OH + H_2O$$

$$NH_2OH + H_2O \longrightarrow NO_2^- + 5\ H^+ + 4\ e^-$$

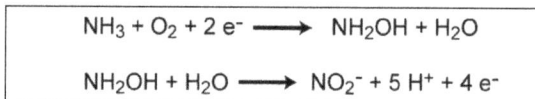

Chemical reactions of ammonia oxidation carried out by bacteria.

Reaction first converts ammonia to the intermediate, hydroxylamine, and is catalyzed by the enzyme ammonia monooxygenase. Reaction second converts hydroxylamine to nitrite and is catalyzed by the enyzmer hydroxylamine oxidoreductase.

Unlike nitrogen fixation that is carried out by many different kinds of microbes, ammonia oxidation is less broadly distributed among prokaryotes. Until recently, it was thought that all ammonia oxidation was carried out by only a few types of bacteria in the genera Nitrosomonas, Nitrosospira, and Nitrosococcus. However, in 2005 an archaeon was discovered that could also oxidize ammonia. Since their discovery, ammonia-oxidizing Archaea have often been found to outnumber the ammonia-oxidizing Bacteria in many habitats. In the past several years, ammonia-oxidizing Archaea have been found to be abundant in oceans, soils, and salt marshes, suggesting an important role in the nitrogen cycle for these newly-discovered organisms. Currently, only one ammonia-oxidizing archaeon has been grown in pure culture, Nitrosopumilus maritimus, so our understanding of their physiological diversity is limited.

The second step in nitrification is the oxidation of nitrite (NO_2^-) to nitrate (NO_3^-). This step is carried out by a completely separate group of prokaryotes, known as nitrite-oxidizing Bacteria. Some of the genera involved in nitrite oxidation include Nitrospira, Nitrobacter, Nitrococcus, and Nitrospina. Similar to ammonia oxidizers, the energy generated from the oxidation of nitrite to nitrate is very small, and thus growth yields are very low. In fact, ammonia- and nitrite-oxidizers must oxidize many molecules of ammonia or nitrite in order to fix a single molecule of CO_2. For complete nitrification, both ammonia oxidation and nitrite oxidation must occur.

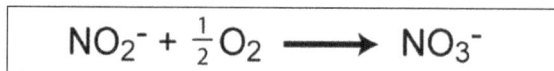

$$NO_2^- + \tfrac{1}{2}O_2 \longrightarrow NO_3^-$$

Chemical reaction of nitrite oxidation.

Ammonia-oxidizers and nitrite-oxidizers are ubiquitous in aerobic environments. They have been extensively studied in natural environments such as soils, estuaries, lakes, and open-ocean environments. However, ammonia- and nitrite-oxidizers also play a very important role in wastewater treatment facilities by removing potentially harmful levels of ammonium that could lead to the pollution of the receiving waters. Much research has focused on how to maintain stable populations of these important microbes in wastewater treatment plants. Additionally, ammonia- and nitrite-oxidizers help to maintain healthy aquaria by facilitating the removal of potentially toxic ammonium excreted in fish urine.

Anammox

Traditionally, all nitrification was thought to be carried out under aerobic conditions, but recently a new type of ammonia oxidation occurring under anoxic conditions was discovered. Anammox (anaerobic ammonia oxidation) is carried out by prokaryotes belonging to the Planctomycetes phylum of Bacteria. The first described anammox bacterium was Brocadia anammoxidans. Anammox bacteria oxidize ammonia by using nitrite as the electron acceptor to produce gaseous nitrogen. Anammox bacteria were first discovered in anoxic bioreactors of wasterwater treatment plants but have since been found in a variety of aquatic systems, including low-oxygen zones of the ocean, coastal and estuarine sediments, mangroves, and freshwater lakes. In some areas of the ocean, the anammox process is considered to be responsible for a significant loss of nitrogen. However, Ward et al. argue that denitrification rather than anammox is responsible for most nitrogen loss in other areas. Whether anammox or denitrification is responsible for most nitrogen loss in the ocean, it is clear that anammox represents an important process in the global nitrogen cycle.

$$NH_4^+ + NO_2^- \longrightarrow N_2 + 2\,H_2O$$

Chemical reaction of anaerobic ammonia oxidation (anammox).

Denitrification

Denitrification is the process that converts nitrate to nitrogen gas, thus removing bio-available nitrogen and returning it to the atmosphere. Dinitrogen gas (N_2) is the ultimate end product of denitrification, but other intermediate gaseous forms of nitrogen exist. Some of these gases, such as nitrous oxide (N_2O), are considered greenhouse gasses, reacting with ozone and contributing to air pollution.

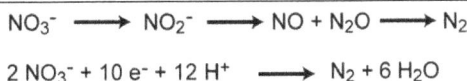

$$NO_3^- \longrightarrow NO_2^- \longrightarrow NO + N_2O \longrightarrow N_2$$
$$2\,NO_3^- + 10\,e^- + 12\,H^+ \longrightarrow N_2 + 6\,H_2O$$

Reactions involved in denitrification-Reaction first represents the steps of reducing nitrate to dinitrogen gas. Reaction second represents the complete redox reaction of denitrification.

Unlike nitrification, denitrification is an anaerobic process, occurring mostly in soils and sediments and anoxic zones in lakes and oceans. Similar to nitrogen fixation, denitrification is carried out by a diverse group of prokaryotes, and there is recent evidence that some eukaryotes are also capable of denitrification. Some denitrifying bacteria include species in the genera Bacillus, Paracoccus, and Pseudomonas. Denitrifiers are chemoorganotrophs and thus must also be supplied with some form of organic carbon.

Denitrification is important in that it removes fixed nitrogen (i.e., nitrate) from the ecosystem and returns it to the atmosphere in a biologically inert form (N_2). This is particularly important in agriculture where the loss of nitrates in fertilizer is detrimental and costly. However, denitrification in wastewater treatment plays a very beneficial role by removing unwanted nitrates from the wastewater effluent, thereby reducing the chances that the water discharged from the treatment plants will cause undesirable consequences (e.g., algal blooms).

Ammonification

When an organism excretes waste or dies, the nitrogen in its tissues is in the form of organic nitrogen (e.g. amino acids, DNA). Various fungi and prokaryotes then decompose the tissue and release inorganic nitrogen back into the ecosystem as ammonia in the process known as ammonification. The ammonia then becomes available for uptake by plants and other microorganisms for growth.

Ecological Implications of Human Alterations to the Nitrogen Cycle

Many human activities have a significant impact on the nitrogen cycle. Burning fossil fuels, application of nitrogen-based fertilizers, and other activities can dramatically increase the amount of biologically available nitrogen in an ecosystem. And because nitrogen availability often limits the primary productivity of many ecosystems, large changes in the availability of nitrogen can lead to severe alterations of the nitrogen cycle in both aquatic and terrestrial ecosystems. Industrial nitrogen fixation has increased exponentially since the 1940s, and human activity has doubled the amount of global nitrogen fixation.

In terrestrial ecosystems, the addition of nitrogen can lead to nutrient imbalance in trees, changes in forest health, and declines in biodiversity. With increased nitrogen availability there is often a change in carbon storage, thus impacting more processes than just the nitrogen cycle. In agricultural systems, fertilizers are used extensively to increase plant production, but unused nitrogen, usually in the form of nitrate, can leach out of the soil, enter streams and rivers, and ultimately make its way into our drinking water. The process of making synthetic fertilizers for use in agriculture by causing N_2 to react with H_2, known as the Haber-Bosch process, has increased significantly over the past several decades. In fact, today, nearly 80% of the nitrogen found in human tissues originated from the Haber-Bosch process.

Much of the nitrogen applied to agricultural and urban areas ultimately enters rivers and nearshore coastal systems. In nearshore marine systems, increases in nitrogen can often lead to anoxia (no oxygen) or hypoxia (low oxygen), altered biodiversity, changes in food-web structure, and general habitat degradation. One common consequence of increased nitrogen is an increase in harmful algal blooms. Toxic blooms of certain types of dinoflagellates have been associated with high fish and shellfish mortality in some areas. Even without such economically catastrophic effects, the addition of nitrogen can lead to changes in biodiversity and species composition that may lead to changes in overall ecosystem function. Some have even suggested that alterations to the nitrogen cycle may lead to an increased risk of parasitic and infectious diseases among humans and wildlife. Additionally, increases in nitrogen in aquatic systems can lead to increased acidification in freshwater ecosystems.

Sulfur Cycle

Sulfur (S), the tenth most abundant element in the universe, is a brittle, yellow, tasteless, and odorless non-metallic element. It comprises many vitamins, proteins, and hormones that play critical roles in both climate and in the health of various ecosystems. The majority of the Earth's sulfur is stored underground in rocks and minerals, including as sulfate salts buried deep within ocean sediments.

The sulfur cycle contains both atmospheric and terrestrial processes. Within the terrestrial portion, the cycle begins with the weathering of rocks, releasing the stored sulfur. The sulfur then comes into contact with air where it is converted into sulfate (SO_4). The sulfate is taken up by plants and microorganisms and is converted into organic forms; animals then consume these organic forms through foods they eat, thereby moving the sulfur through the food chain. As organisms die and decompose, some of the sulfur is again released as a sulfate and some enters the tissues of microorganisms. There are also a variety of natural sources that emit sulfur directly into the atmosphere, including volcanic eruptions, the breakdown of organic matter in swamps and tidal flats, and the evaporation of water.

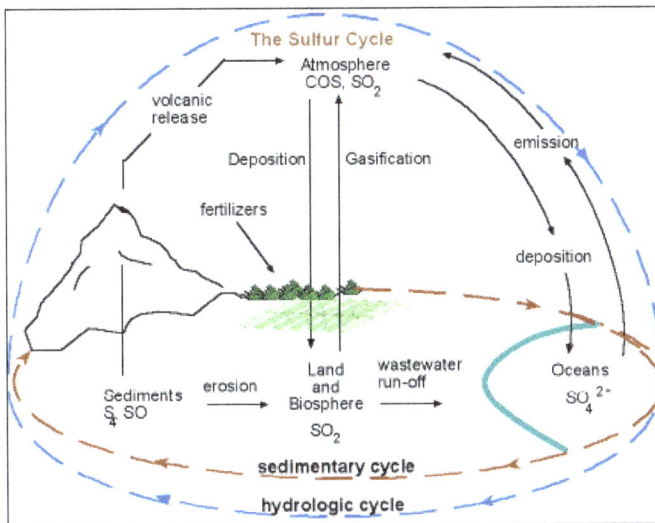

The Sulfur Cycle

Sulfur eventually settles back into the Earth or comes down within rainfall. A continuous loss of sulfur from terrestrial ecosystem runoff occurs through drainage into lakes and streams, and eventually oceans. Sulfur also enters the ocean through fallout from the Earth's atmosphere. Within the ocean, some sulfur cycles through marine communities, moving through the food chain. A portion of this sulfur is emitted back into the atmosphere from sea spray. The remaining sulfur is lost to the ocean depths, combining with iron to form ferrous sulfide which is responsible for the black color of most marine sediments.

Since the Industrial Revolution, human activities have contributed to the amount of sulfur that enters the atmosphere, primarily through the burning of fossil fuels and the processing of metals. One-third of all sulfur that reaches the atmosphere—including 90% of sulfur dioxide—stems from human activities. Emissions from these activities, along with nitrogen emissions, react with other chemicals in the atmosphere to produce tiny particles of sulfate salts which fall as acid rain, causing a variety of damage to both the natural environment as well as to man-made environments, such as the chemical weathering of buildings. However, as particles and tiny airborne droplets, sulfur also acts as a regulator of global climate. Sulfur dioxide and sulfate aerosols absorb ultraviolet radiation, creating cloud cover that cools cities and may offset global warming caused by the greenhouse effect.

Humans and the Sulfur Cycle

Human activities influence the rates and character of certain aspects of the sulfur cycle in important ways, sometimes causing substantial environmental damages.

Acid rain is a well-known environmental problem. Acid rain is ultimately associated with large emissions of sulfur dioxide to the atmosphere by human sources, such as oil- and coal-fired power plants, metal smelters, and the burning of fuel oil to heat homes. The SO_2 is eventually oxidized in the atmosphere to sulfate, much of which is balanced by hydrogen ions, so the precipitation chemistry is acidic. In addition, the vicinity of large point-sources of SO_2 emission is generally polluted by relatively large concentrations of this gas. If its concentration is large enough, the SO_2 can cause toxicity to plants, which may be killed, resulting in severe ecological damages. In addition, atmospheric SO_2 can be directly deposited to surfaces, especially moist soil, plant, or aquatic surfaces, since SO_2 can readily dissolve in water. When this happens, the SO_2 becomes oxidized to sulfate, generating acidity. This means a direct input of sulfur dioxide is called dry deposition, and is a fundamentally different process from the so-called wet deposition of sulfate and acidity with precipitation.

Acid mine drainage is another severe environmental problem that is commonly associated with coal and metal mining, and sometimes with construction activities such as road building. In all of these cases, physical disturbance results in the exposure of large quantities of mineral sulfides to atmospheric oxygen. This causes the sulfides to be oxidized to sulfate, a process accompanied by the generation of large amounts of acidity. Surface waters exposed to acid mine drainage can become severely acidified, to a pH less than 3, resulting in severe biological damages and environmental degradation.

Sulfur is also an important mineral commodity, with many industrial uses in manufacturing. Sulfur for these purposes is largely obtained by cleaning sour natural gas of its content of H_2S, and from pollution control at some metal smelters.

In a few types of intensively managed agriculture, crops may be well fertilized with nitrogen, phosphorus, and other nutrients, and in such cases there may be a deficiency of sulfate availability. Because sulfate is an important plant nutrient, it may have to be applied in the form of a sulfate-containing fertilizer. In North America, sulfate fertilization is most common in prairie agriculture.

Phosphorus Cycle

Phosphorus is an important element for all forms of life. As phosphate (PO_4), it makes up an important part of the structural framework that holds DNA and RNA together. Phosphates are also a critical component of ATP—the cellular energy carrier—as they serve as an energy release' for organisms to use in building proteins or contacting muscles. Like calcium, phosphorus is important to vertebrates; in the human body, 80% of phosphorous is found in teeth and bones.

The phosphorus cycle differs from the other major biogeochemical cycles in that it does not include a gas phase; although small amounts of phosphoric acid (H_3PO_4) may make their way into the atmosphere, contributing—in some cases—to acid rain. The water, carbon, nitrogen and sulfur cycles all include at least one phase in which the element is in its gaseous state. Very little phosphorus circulates in the atmosphere because at Earth's normal temperatures and pressures, phosphorus and its various compounds are not gases. The largest reservoir of phosphorus is in sedimentary rock.

The phosphorus cycle is the process by which phosphorus moves through the lithosphere, hydrosphere, and biosphere. Phosphorus is essential for plant and animal growth, as well as the health of microbes inhabiting the soil, but is gradually depleted from the soil over time. The main biological function of phosphorus is that it is required for the formation of nucleotides, which comprise DNA and RNA molecules. Specifically, the DNA double helix is linked by a phosphate ester bond. Calcium phosphate is also the primary component of mammalian bones and teeth, insect exoskeletons, phospholipid membranes of cells, and is used in a variety of other biological functions. The phosphorus cycle is an extremely slow process, as various weather conditions (e.g., rain and erosion) help to wash the phosphorus found in rocks into the soil. In the soil, the organic matter (e.g., plants and fungi) absorb the phosphorus to be used for various biological processes.

Phosphorus Cycle Steps

The phosphorus cycle is a slow process, which involves five key steps, as shown in the diagram and described as follows:

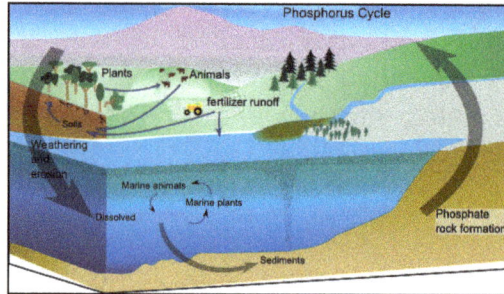

Weathering

Since the main source of phosphorus is found in rocks, the first step of the phosphorus cycle involves the extraction of phosphorus from the rocks by weathering. Weather events, such as rain and other sources of erosion, result in phosphorus being washed into the soil.

Absorption by Plants and Animals

Once in the soil, plants, fungi, and microorganisms are able to absorb phosphorus and grow. In addition, phosphorus can also be washed into the local water systems. Plants can also directly absorb phosphorus from the water and grow. In addition to plants, animals also obtain phosphorus from drinking water and eating plants.

Return to the Environment via Decomposition

When plants and animals die, decomposition results in the return of phosphorus back to the environment via the water or soil. Plants and animals in these environments can then use this phosphorus, and step 2 of the cycle is repeated.

Human Impact on the Phosphorus Cycle

Humans have had a significant impact on the phosphorus cycle due to a variety of human activities, such as the use of fertilizer, the distribution of food products, and artificial eutrophication. Fertilizers containing phosphorus add to the phosphorus levels in the soil and are particularly detrimental when such products are washed into local aquatic ecosystems. When phosphorus is added to waters at a rate typically achieved by natural processes, it is referred to as natural eutrophication. A natural supply of phosphorus over time provides nutrients to the water and serves to increase the productivity of that particular ecosystem. However, when foods are shipped from farms to cities, the substantial levels of Phosphorus that is drained into the water systems is called artificial or anthropogenic eutrophication. When levels of phosphorus are too high, the overabundance of plant nutrients serves to drive the excessive growth of algae. However, these algae die or form algae blooms, which are toxic to the plants and animals in the ecosystem. Thus, human activities serve to harm aquatic ecosystems, whenever excess amounts of phosphorus are leached into the water.

Rock Cycle

There are three types of rocks: igneous, sedimentary and metamorphic. Each of these types is part of the rock cycle. Through changes in conditions one rock type can become another rock type. Or it can become a different rock of the same type.

A rock is a naturally formed, non-living earth material. Rocks are made of collections of mineral grains that are held together in a firm, solid mass.

Figure: The different colors and textures seen in this rock are caused by the presence of different minerals.

How is a rock different from a mineral? Rocks are made of minerals. The mineral grains in a rock may be so tiny that you can only see them with a microscope, or they may be as big as your fingernail or even your finger.

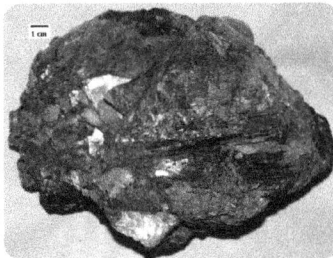

Figure: A pegmatite from South Dakota with crystals of lepidolite, tourmaline, and quartz (1 cm scale on the upper left).

Rocks are identified primarily by the minerals they contain and by their texture. Each type of rock has a distinctive set of minerals. A rock may be made of grains of all one mineral type, such as quartzite. Much more commonly, rocks are made of a mixture of different minerals. Texture is a description of the size, shape, and arrangement of mineral grains. Are the two samples in figure the same rock type? Do they have the same minerals? The same texture?

Rock samples.

Sample	Minerals	Texture	Formation	Rock type
Sample 1	Plagioclase, quartz, hornblende, pyroxene	Crystals, visible to naked eye	Magma cooled slowly	Diorite
Sample 2	Plagioclase, hornblende, pyroxene	Crystals are tiny or microscopic	Magma erupted and cooled quickly	Andesite

As shown in the table above, these two rocks have the same chemical composition and contain mostly the same minerals, but they do not have the same texture. Sample 1 has visible mineral grains, but Sample 2 has very tiny or invisible grains. The two different textures indicate different histories. Sample 1 is a diorite, a rock that cooled slowly from magma (molten rock) underground. Sample 2 is an andesite, a rock that cooled rapidly from a very similar magma that erupted onto Earth's surface.

Three Main Categories of Rocks

Rocks are classified into three major groups according to how they form. Rocks can be studied in hand samples that can be moved from their original location. Rocks can also be studied in outcrop, exposed rock formations that are attached to the ground, at the location where they are found.

Igneous Rocks

This flowing lava is molten rock that will harden into an igneous rock.

Igneous rocks form from cooling magma. Magma that erupts onto Earth's surface is lava, as seen in figure. The chemical composition of the magma and the rate at which it cools determine what rock forms as the minerals cool and crystallize.

Sedimentary Rocks

Sedimentary rocks form by the compaction and cementing together of sediments, broken pieces of rock-like gravel, sand, silt, or clay. Those sediments can be formed from the weathering and erosion of preexisting rocks. Sedimentary rocks also include chemical precipitates, the solid materials left behind after a liquid evaporates.

This sedimentary rock is made of sand that is cemented together to form a sandstone.

Metamorphic Rocks

Metamorphic rocks form when the minerals in an existing rock are changed by heat or pressure within the Earth.

Quartzite is a metamorphic rock formed when quartz sandstone is exposed to heat and pressure within the Earth.

The Rock Cycle

James Hutton is considered the Father of Geology.

Rocks change as a result of natural processes that are taking place all the time. Most changes happen very slowly; many take place below the Earth's surface, so we may not even notice the changes. Although we may not see the changes, the physical and chemical properties of rocks are constantly changing in a natural, never-ending cycle called the rock cycle.

The concept of the rock cycle was first developed by James Hutton, an eighteenth century scientist often called the "Father of Geology". Hutton recognized that geologic processes have "no sign of a beginning, and no prospect of an end." The processes involved in the rock cycle often take place over millions of years. So on the scale of a human lifetime, rocks appear to be "rock solid" and unchanging, but in the longer term, change is always taking place.

In the rock cycle, illustrated in figure, the three main rock types—igneous, sedimentary, and metamorphic—are shown. Arrows connecting the three rock types show the processes that change one rock type into another. The cycle has no beginning and no end. Rocks deep within the Earth are right now becoming other types of rocks. Rocks at the surface are lying in place before they are next exposed to a process that will change them.

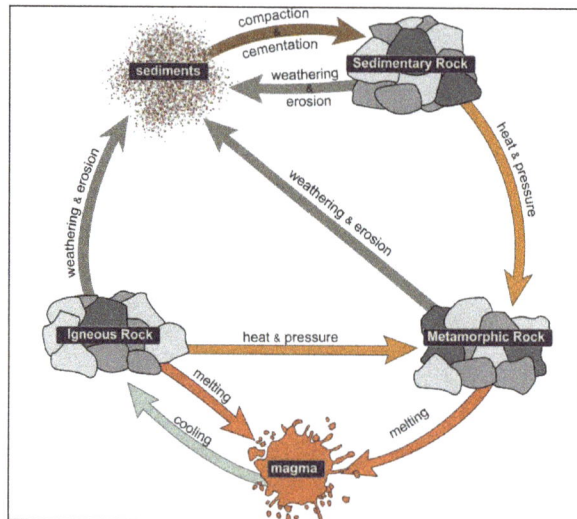

Figure: The Rock Cycle.

Processes of the Rock Cycle

Several processes can turn one type of rock into another type of rock. The key processes of the rock cycle are crystallization, erosion and sedimentation, and metamorphism.

Crystallization

Magma cools either underground or on the surface and hardens into an igneous rock. As the magma cools, different crystals form at different temperatures, undergoing crystallization. For example, the mineral olivine crystallizes out of magma at much higher temperatures than quartz. The rate of cooling determines how much time the crystals will have to form. Slow cooling produces larger crystals.

Erosion and Sedimentation

Weathering wears rocks at the Earth's surface down into smaller pieces. The small fragments are called sediments. Running water, ice, and gravity all transport these sediments from one place to another by erosion. During sedimentation, the sediments are laid down or deposited. In order to form a sedimentary rock, the accumulated sediment must become compacted and cemented together.

Metamorphism

When a rock is exposed to extreme heat and pressure within the Earth but does not melt, the rock becomes metamorphosed. Metamorphism may change the mineral composition and the texture of the rock. For that reason, a metamorphic rock may have a new mineral composition and/or texture.

References

- Earth-cycles, why-is-the-earth-habitable, planet-earth: amnh.org, Retrieved 26 April, 2019

- Oxygen-cycle: universetoday.com, Retrieved 20 August, 2019

- Carbon-cycle, biogeochemical-cycles, air-climate-weather: enviroliteracy.org, Retrieved 11 January, 2019

- Carbon-cycle: biologydictionary.net, Retrieved 21 May, 2019

- The-nitrogen-cycle-processes-players-and-human-15644632: nature.com, Retrieved 13 February, 2019

- Sulfur-cycle, biogeochemical-cycles, air-climate-weather: enviroliteracy.org, Retrieved 14 June, 2019

- Sulfur-Cycle-Humans-sulfur-cycle: jrank.org, Retrieved 9 May, 2019

- Phosphorus-cycle, biogeochemical-cycles, air-climate-weather: enviroliteracy.org, Retrieved 7 March, 2019

- Phosphorus-cycle: biologydictionary.net, Retrieved 6 July, 2019

- Reading-the-rock-cycle, geology: lumenlearning.com, Retrieved 1 April, 2019

Ecosystems on Earth

An ecosystem consists of all the living things in an area such as plants, animals and other organisms. The major types of ecosystems are aquatic ecosystems and terrestrial ecosystems. This chapter closely examines these primary ecosystems to provide an extensive understanding of the subject.

An ecosystem, a term very often used in biology, is a community of plants and animals interacting with each other in a given area, and also with their non-living environments. The non-living environments include weather, earth, sun, soil, climate and atmosphere. The ecosystem relates to the way that all these different organisms live in close proximity to each other and how they interact with each other. For instance, in an ecosystem where there are both rabbits and foxes, these two creatures are in a relationship where the fox eats the rabbit in order to survive. This relationship has a knock on effect with the other creatures and plants that live in the same or similar areas. For instance, the more rabbits that foxes eat, the more the plants may start to thrive because there are fewer rabbits to eat them.

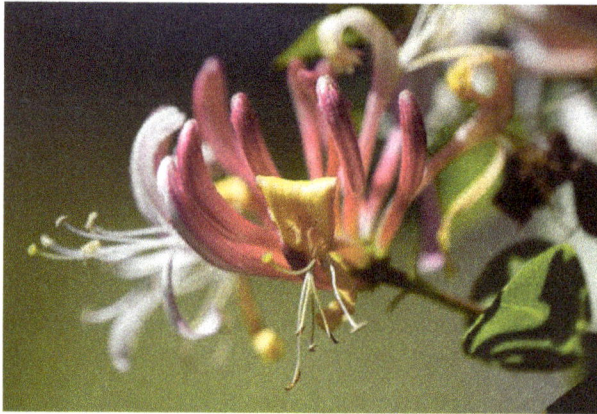

Ecosystems can be huge, with many hundreds of different animals and plants all living in a delicate balance, or they could be relatively small. In particularly harsh places in the world, particularly the North and South Poles, the ecosystems are relatively simple because there are only a few types of creatures that can withstand the freezing temperatures and harsh living conditions. Some creatures can be found in multiple different ecosystems all over the world in different relationships with other or similar creatures. Ecosystems also consist of creatures that mutually benefit from each other. For instance, a popular example is that of the clown fish and the anemone – the clown fish cleans the anemone and keeps it safe from parasites as the anemone stings bigger predators that would otherwise eat clown fish.

An ecosystem can be destroyed by a stranger. The stranger could be rise in temperature or rise in sea level or climate change. The stranger can affect the natural balance and can harm or destroy the ecosystem. It is a bit unfortunate but ecosystems have been destroyed and vanished by man-made activities like deforestation, urbanization and natural activities like floods, storms, fires or volcanic eruptions.

Ecosystem Structure

At a basic functional level, ecosystem generally contains primary producers (plants) capable of harvesting energy from the sun through the process called photosynthesis. This energy then flows through the food chain. Next come consumers. Consumers could be primary consumers (herbivores) or secondary consumers (carnivores). These consumers feed on the captured energy. Decomposers work at the bottom of the food chain. Dead tissues and waste products are produced at all levels. Scavengers, detritivores and decomposers not only feed on this energy but also break organic matter back into its organic constituents. It is the microbes that finish the job of decomposition and produce organic constituents that can again be used by producers.

Energy that flows through the food chain i.e. from producers to consumers to decomposers is always inefficient. That means less energy is available at secondary consumers level than at primary producers level. It is not surprising but amount of energy produced from place to place varies a lot due to amount of solar radiation and the availability of nutrients and water.

Aquatic Ecosystems

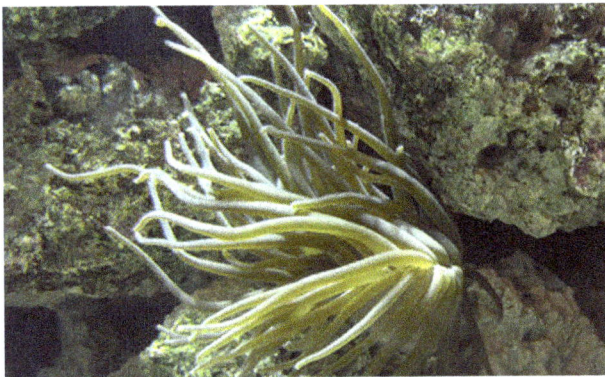

Aquatic ecosystems are any water-based environment in which plants and animals interact with the chemical and physical features of the aquatic environment. Aquatic ecosystems are generally divided into two types --the marine ecosystem and the fresh-water ecosystem. The largest water ecosystem is the marine ecosystem, covering over 70 percent of the earth's surface. Oceans, estuaries, coral reefs and coastal ecosystems are the various kinds of marine ecosystems. Freshwater ecosystems cover less than 1 percent of the earth and are subdivided into lotic, lentic and wetlands.

Marine Ecosystems

Marine ecosystem is complex of living organisms in the ocean environment. Marine waters cover two-thirds of the surface of the Earth. In some places the ocean is deeper than Mount Everest is high; for example, the Mariana Trench and the Tonga Trench in the western part of the Pacific Ocean reach depths in excess of 10,000 metres (32,800 feet). Within this ocean habitat live a wide variety of organisms that have evolved in response to various features of their environs.

The Marine Environment

Geography, Oceanography and Topography

The shape of the oceans and seas of the world has changed significantly throughout the past 600 million years. According to the theory of plate tectonics, the crust of the Earth is made up of many dynamic plates. There are two types of plates—oceanic and continental—which float on the surface of the Earth's mantle, diverging, converging, or sliding against one another. When two plates diverge, magma from the mantle wells up and cools, forming new crust; when convergence occurs, one plate descends—i.e., is subducted—below the other and crust is resorbed into the mantle. Examples of both processes are observed in the marine environment. Oceanic crust is created along oceanic ridges or rift areas, which are vast undersea mountain ranges such as the Mid-Atlantic Ridge. Excess crust is reabsorbed along subduction zones, which usually are marked by deep-sea trenches such as the Kuril Trench off the coast of Japan.

The shape of the ocean also is altered as sea levels change. During ice ages a higher proportion of the waters of the Earth is bound in the polar ice caps, resulting in a relatively low sea level. When the polar ice caps melt during interglacial periods, the sea level rises. These changes in sea level cause great changes in the distribution of marine environments such as coral reefs. For example, during the last Pleistocene Ice Age the Great Barrier Reef did not exist as it does today; the continental shelf on which the reef now is found was above the high-tide mark.

Marine organisms are not distributed evenly throughout the oceans. Variations in characteristics of the marine environment create different habitats and influence what types of organisms will inhabit them. The availability of light, water depth, proximity to land, and topographic complexity all affect marine habitats.

The availability of light affects which organisms can inhabit a certain area of a marine ecosystem. The greater the depth of the water, the less light can penetrate until below a certain depth there is no light whatsoever. This area of inky darkness, which occupies the great bulk of the ocean, is called the aphotic zone. The illuminated region above it is called the photic zone, within which are distinguished the euphotic and disphotic zones. The euphotic zone is the layer closer to the surface that receives enough light for photosynthesis to occur. Beneath lies the disphotic zone, which is illuminated but so poorly that rates of respiration exceed those of photosynthesis. The actual depth of these zones depends on local conditions of cloud cover, water turbidity, and ocean surface. In general, the euphotic zone can extend to depths of 80 to 100 metres and the disphotic zone to depths of 80 to 700 metres. Marine organisms are particularly abundant in the photic zone, especially the euphotic portion; however, many organisms inhabit the aphotic zone and migrate vertically to the photic zone every night. Other organisms, such as the tripod fish and some species of sea cucumbers and brittle stars, remain in darkness all their lives.

Marine environments can be characterized broadly as a water, or pelagic, environment and a bottom, or benthic, environment. Within the pelagic environment the waters are divided into the neritic province, which includes the water above the continental shelf, and the oceanic province, which includes all the open waters beyond the continental shelf. The high nutrient levels of the neritic province resulting from dissolved materials in riverine runoff distinguish this province from the oceanic. The upper portion of both the neritic and oceanic waters—the epipelagic zone—is where photosynthesis occurs; it is roughly equivalent to the photic zone. Below this zone lie the mesopelagic, ranging between 200 and 1,000 metres, the bathypelagic, from 1,000 to 4,000 metres, and the abyssalpelagic, which encompasses the deepest parts of the oceans from 4,000 metres to the recesses of the deep-sea trenches.

The benthic environment also is divided into different zones. The supralittoral is above the high-tide mark and is usually not under water. The intertidal, or littoral, zone ranges from the high-tide mark (the maximum elevation of the tide) to the shallow, offshore waters. The sublittoral is the environment beyond the low-tide mark and is often used to refer to substrata of the continental shelf, which reaches depths of between 150 and 300 metres. Sediments of the continental shelf that influence marine organisms generally originate from the land, particularly in the form of riverine runoff, and include clay, silt, and sand. Beyond the continental shelf is the bathyal zone, which occurs at depths of 150 to 4,000 metres and includes the descending continental slope and rise. The abyssal zone (between 4,000 and 6,000 metres) represents a substantial portion of the oceans. The deepest region of the oceans (greater than 6,000 metres) is the hadal zone of the deep-sea trenches. Sediments of the deep sea primarily originate from a rain of dead marine organisms and their wastes.

Physical and Chemical Properties of Seawater

The physical and chemical properties of seawater vary according to latitude, depth, nearness to land, and input of fresh water. Approximately 3.5 percent of seawater is composed of dissolved compounds, while the other 96.5 percent is pure water. The chemical composition of seawater reflects such processes as erosion of rock and sediments, volcanic activity, gas exchange with the atmosphere, the metabolic and breakdown products of organisms, and rain. (For a list of the principal constituents of seawater,) In addition to carbon, the nutrients essential for living organisms include nitrogen and phosphorus, which are minor constituents of seawater and thus are often limiting factors in organic cycles of the ocean. Concentrations of phosphorus and nitrogen are generally low in the photic zone because they are rapidly taken up by marine organisms. The highest concentrations of these nutrients generally are found below 500 metres, a result of the decay of organisms. Other important elements include silicon (used in the skeletons of radiolarians and diatoms;) and calcium (essential in the skeletons of many organisms such as fish and corals).

The chemical composition of the atmosphere also affects that of the ocean. For example, carbon dioxide is absorbed by the ocean and oxygen is released to the atmosphere through the activities of marine plants. The dumping of pollutants into the sea also can affect the chemical makeup of the ocean, contrary to earlier assumptions that, for example, toxins could be safely disposed of there.

The physical and chemical properties of seawater have a great effect on organisms, varying especially with the size of the creature. As an example, seawater is viscous to very small animals (less than 1 millimetre [0.039 inch] long) such as ciliates but not to large marine creatures such as tuna.

Marine organisms have evolved a wide variety of unique physiological and morphological features that allow them to live in the sea. Notothenid fishes in Antarctica are able to inhabit waters as cold as –2 °C (28 °F) because of proteins in their blood that act as antifreeze. Many organisms are able to achieve neutral buoyancy by secreting gas into internal chambers, as cephalopods do, or into swim bladders, as some fish do; other organisms use lipids, which are less dense than water, to achieve this effect. Some animals, especially those in the aphotic zone, generate light to attract prey. Animals in the disphotic zone such as hatchet fish produce light by means of organs called photophores to break up the silhouette of their bodies and avoid visual detection by predators. Many marine animals can detect vibrations or sound in the water over great distances by means of specialized organs. Certain fishes have lateral line systems, which they use to detect prey, and whales have a sound-producing organ called a melon with which they communicate. Tolerance to differences in salinity varies greatly: stenohaline organisms have a low tolerance to salinity changes, whereas euryhaline organisms, which are found in areas where river and sea meet (estuaries), are very tolerant of large changes in salinity. Euryhaline organisms are also very tolerant of changes in temperature. Animals that migrate between fresh water and salt water, such as salmon or eels, are capable of controlling their osmotic environment by active pumping or the retention of salts. Body architecture varies greatly in marine waters. The body shape of the cnidarian by-the-wind-sailor (Velella velella)—an animal that lives on the surface of the water (pleuston) and sails with the assistance of a modified flotation chamber—contrasts sharply with the sleek, elongated shape of the barracuda.

Ocean Currents

The movements of ocean waters are influenced by numerous factors, including the rotation of the Earth (which is responsible for the Coriolis Effect), atmospheric circulation patterns that influence surface waters, and temperature and salinity gradients between the tropics and the polar regions (thermohaline circulation). The resultant patterns of circulation range from those that cover great areas, such as the North Subtropical Gyre, which follows a path thousands of kilometres long, to small-scale turbulences of less than one metre.

Marine organisms of all sizes are influenced by these patterns, which can determine the range of a species. For example, krill (Euphausia superba) are restricted to the Antarctic Circumpolar Current. Distribution patterns of both large and small pelagic organisms are affected as well. Mainstream currents such as the Gulf Stream and East Australian Current transport larvae great distances. As a result cold temperate coral reefs receive a tropical infusion when fish and invertebrate larvae from the tropics are relocated to high latitudes by these currents. The successful recruitment of eels to Europe depends on the strength of the Gulf Stream to transport them from spawning sites in the Caribbean. Areas where the ocean is affected by nearshore features, such as estuaries, or areas in which there is a vertical salinity gradient (halocline) often exhibit intense biological activity. In these environments, small organisms can become concentrated, providing a rich supply of food for other animals.

Marine Biota

Marine biota can be classified broadly into those organisms living in either the pelagic environment (plankton and nekton) or the benthic environment (benthos). Some organisms, however, are benthic in one stage of life and pelagic in another. Producers that synthesize organic molecules exist in both environments. Single-celled or multicelled plankton with photosynthetic pigments are the producers of the photic zone in the pelagic environment. Typical benthic producers are microalgae (e.g., diatoms), macroalgae (e.g., the kelp Macrocystis pyrifera), or sea grass (e.g., Zostera).

Plankton

Plankton is the numerous, primarily microscopic inhabitants of the pelagic environment. They are critical components of food chains in all marine environments because they provide nutrition for the nekton (e.g., crustaceans, fish, and squid) and benthos (e.g., sea squirts and sponges). They also exert a global effect on the biosphere because the balance of components of the Earth's atmosphere depends to a great extent on the photosynthetic activities of some plankton.

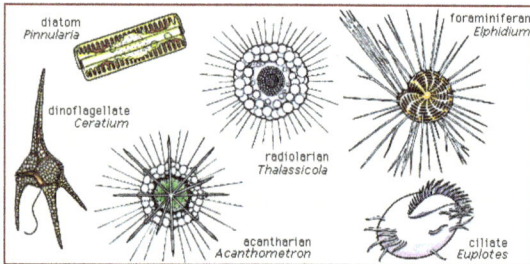

Representative plankton.

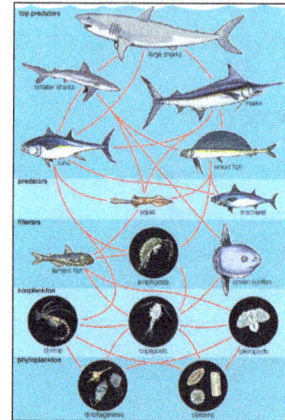

Generalized aquatic food web.

Plankton range in size from tiny microbes (1 micrometre [0.000039 inch] or less) to jellyfish whose gelatinous bell can reach up to 2 metres in width and whose tentacles can extend over 15 metres. However, most planktonic organisms, called plankters, are less than 1 millimetre (0.039 inch) long. These microbes thrive on nutrients in seawater and are often photosynthetic. The plankton include a wide variety of organisms such as algae, bacteria, protozoans, the larvae of some animals, and crustaceans. A large proportion of the plankton are protists—i.e., eukaryotic, predominantly single-celled organisms. Plankton can be broadly divided into phytoplankton, which are plants or plantlike protists; zooplankton, which are animals or animal-like protists; and microbes such as bacteria. Phytoplankton carry out photosynthesis and are the producers of the marine community; zooplankton are the heterotrophic consumers.

Diatoms and dinoflagellates (approximate range between 15 and 1,000 micrometres in length) are two highly diverse groups of photosynthetic protists that are important components of the plankton. Diatoms are the most abundant phytoplankton. While many dinoflagellates carry out photosynthesis, some also consume bacteria or algae. Other important groups of protists include flagellates, foraminiferans, radiolarians, acantharians, and ciliates. Many of these protists are important consumers and a food source for zooplankton.

Zooplankton which are greater than 0.05 millimetre in size, are divided into two general categories: meroplankton, which spend only a part of their life cycle—usually the larval or juvenile stage—as plankton, and holoplankton, which exist as plankton all their lives. Many larval meroplankton in coastal, oceanic, and even freshwater environments (including sea urchins, intertidal snails, and crabs, lobsters, and fish) bear little or no resemblance to their adult forms. These larvae may exhibit features unique to the larval stage, such as the spectacular spiny armour on the larvae of certain crustaceans (e.g., Squilla), probably used to ward off predators.

Important holoplanktonic animals include such lobsterlike crustaceans as the copepods, cladocerans, and euphausids (krill), which are important components of the marine environment because they serve as food sources for fish and marine mammals. Gelatinous forms such as larvaceans, salps, and siphonophores graze on phytoplankton or other zooplankton. Some omnivorous zooplankton such as euphausids and some copepods consume both phytoplankton and zooplankton; their feeding behaviour changes according to the availability and type of prey. The grazing and predatory activity of some zooplankton can be so intense that measurable reductions in phytoplankton or zooplankton abundance (or biomass) occur. For example, when jellyfish occur in high concentration in enclosed seas, they may consume such large numbers of fish larvae as to greatly reduce fish populations.

The jellylike plankton are numerous and predatory. They secure their prey with stinging cells (nematocysts) or sticky cells (colloblasts of comb jellies). Large numbers of the Portuguese man-of-war (Physalia), with its conspicuous gas bladder, the by-the-wind-sailor (Velella velella), and the small blue disk-shaped Porpita porpita are propelled along the surface by the wind, and after strong onshore winds they may be found strewn on the beach. Beneath the surface, comb jellies often abound, as do siphonophores, salps, and scyphomedusae.

The pelagic environment was once thought to present few distinct habitats, in contrast to the array of niches within the benthic environment. Because of its apparent uniformity, the pelagic realm was understood to be distinguished simply by plankton of different sizes. Small-scale variations in the pelagic environment, however, have been discovered that affect biotic distributions. Living and dead matter form organic aggregates called marine snow to which members of the plankton community may adhere, producing patchiness in biotic distributions. Marine snow includes structures such as aggregates of cells and mucus as well as drifting macroalgae and other flotsam that range in size from 0.5 millimetre to 1 centimetre (although these aggregates can be as small as 0.05 millimetre and as large as 100 centimetres). Many types of microbes, phytoplankton, and zooplankton stick to marine snow, and some grazing copepods and predators will feed from the surface of these structures. Marine snow is extremely abundant at times, particularly after plankton blooms. Significant quantities of organic material from upper layers of the ocean may sink to the ocean floor as marine snow, providing an important source of food for bottom dwellers. Other structures that plankton respond to in the marine environment include aggregates of phytoplankton cells that form large rafts in tropical and temperate waters of the world (e.g., cells of Oscillatoria [Trichodesmium] erthraeus) and various types of seaweed (e.g., Sargassum, Phyllospora, Macrocystis) that detach from the seafloor and drift.

Nekton

Nektons are the active swimmers of the oceans and are often the best-known organisms of marine waters. Nektons are the top predators in most marine food chains. The distinction between nekton and plankton is not always sharp. As mentioned above, many large marine animals, such as marlin and tuna, spend the larval stage of their lives as plankton and their adult stage as large and active members of the nekton. Other organisms such as krill are referred to as both micronekton and macrozooplankton.

The vast majority of nektons are vertebrates (e.g., fishes, reptiles, and mammals), mollusks, and crustaceans. The most numerous group of nekton are the fishes, with approximately 16,000 species. Nektons are found at all depths and latitudes of marine waters. Whales, penguins, seals, and icefish abound in polar waters. Lantern fish (family Myctophidae) are common in the aphotic zone along with gulpers (Saccopharynx), whalefish (family Cetomimidae), seven-gilled sharks, and others. Nekton diversity is greatest in tropical waters, where in particular there are large numbers of fish species.

The largest animals on the Earth, the blue whales (Balaenoptera musculus), which grow to 25 to 30 metres long, are members of the nekton. These huge mammals and other baleen whales (order Mysticeti), which are distinguished by fine filtering plates in their mouths, feed on plankton and micronekton as do whale sharks (Rhinocodon typus), the largest fish in the world (usually 12 to 14 metres long, with some reaching 17 metres). The largest carnivores that consume large prey include the toothed whales (order Odontoceti—for example, the killer whales, Orcinus orca), great white sharks (Carcharodon carcharias), tiger sharks (Galeocerdo cuvier), black marlin (Makaira indica), bluefin tuna (Thunnus thynnus), and giant groupers (Epinephelus lanceolatus).

Nekton forms the basis of important fisheries around the world. Vast schools of small anchovies, herring, and sardines generally account for one-quarter to one-third of the annual harvest from the ocean. Squid are also economically valuable nekton. Halibut, sole, and cod are demersal (i.e., bottom-dwelling) fish that are commercially important as food for humans. They are generally caught in continental shelf waters. Because pelagic nekton often abound in areas of upwelling where the waters are nutrient-rich, these regions also are major fishing areas.

Benthos

Organisms are abundant in surface sediments of the continental shelf and in deeper waters, with a great diversity found in or on sediments. In shallow waters, beds of seagrass provide a rich habitat for polychaete worms, crustaceans (e.g., amphipods), and fishes. On the surface of and within intertidal sediments most animal activities are influenced strongly by the state of the tide. On many sediments in the photic zone, however, the only photosynthetic organisms are microscopic benthic diatoms.

Benthic organisms can be classified according to size. The macrobenthos are those organisms larger than 1 millimetre. Those that eat organic material in sediments are called deposit feeders (e.g., holothurians, echinoids, gastropods), those that feed on the plankton above are the suspension feeders (e.g., bivalves, ophiuroids, crinoids), and those that consume other fauna in the benthic assemblage are predators (e.g., starfish, gastropods). Organisms between 0.1 and 1 millimetre constitute the meiobenthos. These larger microbes, which include foraminiferans, turbellarians, and polychaetes, frequently dominate benthic food chains, filling the roles of nutrient recycler, decomposer, primary producer, and predator. The microbentho are those organisms smaller than 1 millimetre; they include diatoms, bacteria, and ciliates.

Organic matter is decomposed aerobically by bacteria near the surface of the sediment where oxygen is abundant. The consumption of oxygen at this level, however, deprives deeper layers of oxygen, and marine sediments below the surface layer are anaerobic. The thickness of the oxygenated layer varies according to grain size, which determines how permeable the sediment is to oxygen and the amount of organic matter it contains. As oxygen concentration diminishes, anaerobic processes come to dominate. The transition layer between oxygen-rich and oxygen-poor layers is called the redox discontinuity layer and appears as a gray layer above the black anaerobic layers. Organisms have evolved various ways of coping with the lack of oxygen. Some anaerobes release hydrogen sulfide, ammonia, and other toxic reduced ions through metabolic processes. The thiobiota, made up primarily of microorganisms, metabolize sulfur. Most organisms that live below the redox layer, however, have to create an aerobic environment for themselves. Burrowing animals generate a respiratory current along their burrow systems to oxygenate their dwelling places; the influx of oxygen must be constantly maintained because the surrounding anoxic layer quickly depletes the burrow of oxygen. Many bivalves (e.g., Mya arenaria) extend long siphons upward into oxygenated waters near the surface so that they can respire and feed while remaining sheltered from predation deep in the sediment. Many large mollusks use a muscular "foot" to dig with, and in some cases they use it to propel themselves away from predators such as starfish. The consequent "irrigation" of burrow systems can create oxygen and nutrient fluxes that stimulate the production of benthic producers (e.g., diatoms).

Not all benthic organisms live within the sediment; certain benthic assemblages live on a rocky substrate. Various phyla of algae—Rhodophyta (red), Chlorophyta (green), and Phaeophyta (brown)—are abundant and diverse in the photic zone on rocky substrata and are important producers. In intertidal regions algae are most abundant and largest near the low-tide mark. Ephemeral algae such as Ulva, Enteromorpha, and coralline algae cover a broad range of the intertidal. The mix of algae species found in any particular locale is dependent on latitude and also varies greatly according to wave exposure and the activity of grazers. For example, Ascophyllum spores cannot attach to rock in even a gentle ocean surge; as a result this plant is largely restricted to sheltered shores. The fastest-growing plant—adding as much as 1 metre per day to its length—is the giant kelp, Macrocystis pyrifera, which is found on subtidal rocky reefs. These plants, which may exceed 30 metres in length, characterize benthic habitats on many temperate reefs. Large laminarian and fucoid algae are also common on temperate rocky reefs, along with the encrusting (e.g., Lithothamnion) or short tufting forms (e.g., Pterocladia).

Many algae on rocky reefs are harvested for food, fertilizer, and pharmaceuticals. Macroalgae are relatively rare on tropical reefs where corals abound, but Sargassum and a diverse assemblage of short filamentous and tufting algae are found, especially at the reef crest. Sessile and slow-moving invertebrates are common on reefs. In the intertidal and subtidal regions herbivorous gastropods and urchins abound and can have a great influence on the distribution of algae. Barnacles are common sessile animals in the intertidal. In the subtidal regions, sponges, ascidians, urchins, and anemones are particularly common where light levels drop and current speeds are high. Sessile assemblages of animals are often rich and diverse in caves and under boulders.

Reef-building coral polyps (Scleractinia) are organisms of the phylum Cnidaria that create a calcareous substrate upon which a diverse array of organisms live. Approximately 700 species of corals are found in the Pacific and Indian oceans and belong to genera such as Porites, Acropora, and Montipora. Some of the world's most complex ecosystems are found on coral reefs. Zooxanthellae are the photosynthetic, single-celled algae that live symbiotically within the tissue of corals and help to build the solid calcium carbonate matrix of the reef. Reef-building corals are found only in waters warmer than 18 °C; warm temperatures are necessary, along with high light intensity, for the coral-algae complex to secrete calcium carbonate. Many tropical islands are composed entirely of hundreds of metres of coral built atop volcanic rock.

Links between the Pelagic Environments and the Benthos- Considering the pelagic and benthic environments in isolation from each other should be done cautiously because the two are interlinked in many ways. For example, pelagic plankton are an important source of food for animals on soft or rocky bottoms. Suspension feeders such as anemones and barnacles filter living and dead particles from the surrounding water while detritus feeders graze on the accumulation of particulate material raining from the water column above. The molts of crustaceans, plankton feces, dead plankton, and marine snow all contribute to this rain of fallout from the pelagic environment to the ocean bottom. This fallout can be so intense in certain weather patterns—such as the El Niño condition—that benthic animals on soft bottoms are smothered and die. There also is variation in the rate of fallout of the plankton according to seasonal cycles of production. This variation can create seasonality in the abiotic zone where there is little or no variation in temperature or light. Plankton form marine sediments and many types of fossilized protistan plankton, such as foraminiferans and coccoliths are used to determine the age and origin of rocks.

Organisms of the Deep-Sea Vents

Producers were discovered in the aphotic zone when exploration of the deep sea by submarine became common in the 1970s. Deep-sea hydrothermal vents now are known to be relatively common in areas of tectonic activity (e.g., spreading ridges). The vents are a non-photosynthetic source of organic carbon available to organisms. A diversity of deep-sea organisms including mussels, large bivalve clams, and vestimentiferan worms are supported by bacteria that oxidize sulfur (sulfide) and derive chemical energy from the reaction. These organisms are referred to as chemoautotrophic, or chemosynthetic, as opposed to photosynthetic, organisms. Many of the species in the vent fauna have developed symbiotic relationships with chemoautotrophic bacteria, and as a consequence the megafauna are principally responsible for the primary production in the vent assemblage. The situation is analogous to that found on coral reefs where individual coral polyps have symbiotic relationships with zooxanthellae. In addition to symbiotic bacteria there is a rich assemblage of free-living bacteria around vents. For example, Beggiatoas-like bacteria often form conspicuous weblike mats on any hard surface; these mats have been shown to have chemoautotrophic metabolism. Large numbers of brachyuran (e.g., Bythograea) and galatheid crabs, large sea anemones (e.g., Actinostola callasi), copepods, other plankton, and some fish—especially the eelpout Thermarces cerberus—are found in association with vents.

Hydrothermal mussels: Galatheid crabs and shrimp grazing on the bacterial filaments that grow on the shells of the hydrothermal mussels covering the Northwest Eifuku volcano in the Mariana Arc region.

Patterns and Processes influencing the Structure of Marine Assemblages

Distribution and Dispersal

The distribution patterns of marine organisms are influenced by physical and biological processes in both ecological time (tens of years) and geologic time (hundreds to millions of years). The shapes of the Earth's oceans have been influenced by plate tectonics, and as a consequence the distributions of fossil and extant marine organisms also have been affected. Vicariance theory argues that plate tectonics has a major role in determining biogeographic patterns. For example, Australia was once—90 million years ago—close to the South Pole and had few coral reefs. Since then Australia has been moving a few millimetres each year closer to the Equator. As a result of this movement and local oceanographic conditions, coral reef environments are extending ever so slowly southward. Dispersal may also have an important role in biogeographic patterns of abundance. The importance of dispersal varies greatly with local oceanographic features, such as the direction and intensity of currents and the biology of the organisms. Humans can also have an impact on patterns of distribution and the extinction of marine organisms. For example, fishing intensity in the Irish Sea was based on catch limits set for cod with no regard for the biology of other species. One consequence of this practice was that the local skate, which had a slow reproductive rate, was quickly fished to extinction.

A characteristic of many marine organisms is a bipartite life cycle, which can affect the dispersal of an organism. Most animals found on soft and hard substrata, such as lobsters, crabs, barnacles, fish, polychaete worms, and sea urchins, spend their larval phase in the plankton and in this phase are dispersed most widely. The length of the larval phase, which can vary from a few minutes to hundreds of days, has a major influence on dispersal. For example, wrasses of the genus Thalassoma have a long larval life, compared with many other types of reef fish, and populations of these fish are well dispersed to the reefs of isolated volcanic islands around the Pacific. The bipartite life cycle of algae also affects their dispersal, which occurs through algal spores. Although in general, spores disperse only a short distance from adult plants, limited swimming abilities—Macrocystis spores have flagella—and storms can disperse spores over greater distances.

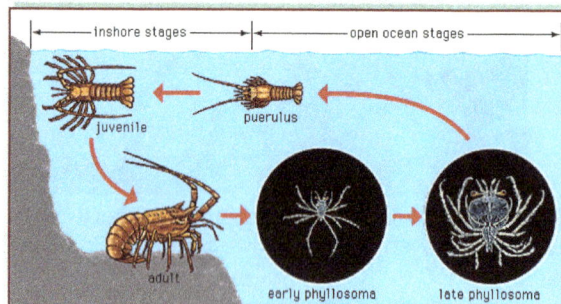

Life cycle of a palinurid lobster.

Migrations of Marine Organisms

The migrations of plankton and nekton throughout the water column in many parts of the world are well described. Diurnal vertical migrations are common. For example, some types of plankton, fish, and squid remain beneath the photic zone during the day, moving toward the surface after dusk and returning to the depths before dawn. It is generally argued that marine organisms migrate in response to light levels. This behaviour may be advantageous because by spending the daylight hours in the dim light or darkness beneath the photic zone plankton can avoid predators that locate their prey visually. After the Sun has set, plankton can rise to the surface waters where food is more abundant and where they can feed safely under the cover of darkness.

Larval forms can facilitate their horizontal transport along different currents by migrating vertically. This is possible because currents can differ in direction according to depth (e.g., above and below haloclines and thermoclines), as is the case in estuaries.

In coastal waters many larger invertebrates (e.g., mysids, amphipods, and polychaete worms) leave the cover of algae and sediments to migrate into the water column at night. It is thought that these animals disperse to different habitats or find mates by swimming when visual predators find it hard to see them. In some cases only one sex will emerge at night, and often that sex is morphologically better suited for swimming.

Horizontal migrations of fish that span distances of hundreds of metres to tens of kilometres are common and generally related to patterns of feeding or reproduction. Tropical coral trout (Plectropomus species) remain dispersed over a reef for most of the year, but adults will aggregate at certain locations at the time of spawning. Transoceanic migrations (greater than 1,000 kilometres) are observed in a number of marine vertebrates, and these movements often relate to requirements of feeding and reproduction. Bluefin tuna (Thunnus thynnus) traverse the Atlantic Ocean in a single year; they spawn in the Caribbean, then swim to high latitudes of the Atlantic to feed on the rich supply of fish. Turtles and sharks also migrate great distances.

Fish that spend their lives in both marine and freshwater systems (diadromous animals) exhibit some of the most spectacular migratory behaviour. Anadromous fishes (those that spend most of their lives in the sea but migrate to fresh water to spawn) such as Atlantic salmon (Salmo salar) also have unique migratory patterns. After spawning, the adults die. Newly hatched fish (alevin) emerge from spawned eggs and develop into young fry that move down rivers toward the sea. Juveniles (parr) grow into larger fish (smolt) that convene near the ocean. When the adult fish are ready to spawn, they return to the river in which they were born (natal river), using a variety of environmental cues, including the Earth's magnetic field, the Sun, and water chemistry. It is believed that the thyroid gland has a role in imprinting the water chemistry of the natal river on the fish. Freshwater eels such as the European eel (Anguilla anguilla) undertake great migrations from fresh water to spawn in the marine waters of the Sargasso Sea (catadromous migrations), where they die. Eel larvae, called leptocephalus larvae, drift back to Europe in the Gulf Stream.

Dynamics of Populations and Assemblages

A wide variety of processes influence the dynamics of marine populations of individual species and the composition of assemblages (e.g., collections of populations of different species that live in the same area). With the exception of marine mammals such as whales, fish that bear live young (e.g., embiotocid fish), and brooders (i.e., fauna that incubate their offspring until they emerge as larvae or juveniles), most marine organisms produce a large number of offspring of which few survive. Processes that affect the plankton can have a great influence on the numbers of young that survive to be recruited, or relocated, into adult populations. The survival of larvae may depend on the abundance of food at various times and in various places, the number of predators, and oceanographic features that retain larvae near suitable nursery areas. The number of organisms recruited to benthic and pelagic systems may ultimately determine the size of adult populations and therefore the relative abundance of species in marine assemblages. However, many processes can affect the survival of organisms after recruitment. Predators eat recruits, and mortality rates in prey species can vary with time and space, thus changing original population patterns established in recruitment.

Patterns of colonization and succession can have a significant impact on benthic assemblages. For example, when intertidal reefs are cleared experimentally, the assemblage of organisms that colonize the bare space often reflects the types of larvae available in local waters at the time. Tube worms may dominate if they establish themselves first; if they fail to do so, algal spores may colonize the shore first and inhibit the settlement of these worms. Competition between organisms may also play a role. Long-term data gathered over periods of more than 25 years from coral reefs have demonstrated that some corals (e.g., Acropora cytherea) competitively overgrow neighbouring corals. Physical disturbance from hurricanes destroys many corals, and during regrowth competitively inferior species can coexist with normally dominant species on the reef. Chemical defenses of sessile organisms also can deter the growth or cause increased mortality of organisms that settle on them. Ascidian larvae (e.g., Podoclavella) often avoid settling on sponges (e.g., Mycale); when this does occur, the larvae rarely reach adulthood.

Although the processes that determine species assemblages may be understood, variations occur in the composition of the plankton that makes it difficult to predict patterns of colonization with great accuracy.

Biological Productivity

Primary productivity is the rate at which energy is converted by photosynthetic and chemosynthetic autotrophs to organic substances. The total amount of productivity in a region or system is gross primary productivity. A certain amount of organic material is used to sustain the life of producers; what remains is net productivity. Net marine primary productivity is the amount of organic material available to support the consumers (herbivores and carnivores) of the sea. The standing crop is the total biomass (weight) of vegetation. Most primary productivity is carried out by pelagic phytoplankton, not benthic plants.

Most primary producers require nitrogen and phosphorus, which are available in the ocean as nitrate, nitrite, ammonia, and phosphorus. The abundances of these molecules and the intensity and quality of light exert a major influence on rates of production. The two principal categories of producers (autotrophs) in the sea are pelagic phytoplankton and benthic microalgae and macroalgae. Benthic plants grow only on the fringe of the world's oceans and are estimated to produce only 5 to 10 percents of the total marine plant material in a year. Chemoautotrophs are the producers of the deep-sea vents.

Primary productivity is usually determined by measuring the uptake of carbon dioxide or the output of oxygen. Production rates are usually expressed as grams of organic carbon per unit area per unit time. The productivity of the entire ocean is estimated to be approximately 16 × 1010 tons of carbon per year, which is about eight times that of the land.

The Pelagic Food Chain

Food chains in coastal waters of the world are generally regulated by nutrient concentrations. These concentrations determine the abundance of phytoplankton, which in turn provide food for the primary consumers, such as protozoa and zooplankton that the higher-level consumers—fish, squid, and marine mammals—prey upon. It had been thought that phytoplankton in the 5- to 100-micrometre size range were responsible for most of the primary production in the sea and that grazers such as copepods controlled the numbers of phytoplankton. Data gathered since 1975, however, indicate that the system is much more complex than this. It is now thought that most primary production in marine waters of the world is accomplished by single-celled 0.5- to 10-micrometre phototrophs (bacteria and protists). Moreover, heterotrophic protists (phagotrophic protists) are now viewed as the dominant controllers of both bacteria and primary production in the sea. Current models of pelagic marine food chains picture complex interactions within a microbial food web. Larger metazoans are supported by the production of autotrophic and heterotrophic cells.

Upwelling

The most productive waters of the world are in regions of upwelling. Upwelling in coastal waters brings nutrients toward the surface. Phytoplankton reproduces rapidly in these conditions, and grazing zooplankton also multiply and provide abundant food supplies for nekton. Some of the world's richest fisheries are found in regions of upwelling—for example, the temperate waters off Peru and California. If upwelling fails, the effects on animals that depend on it can be disastrous. Fisheries also suffer at these times, as evidenced by the collapse of the Peruvian anchovy industry in the 1970s. The intensity and location of upwelling are influenced by changes in atmospheric circulation, as exemplified by the influence of El Niño conditions.

Seasonal Cycles of Production

Cycles of plankton production vary at different latitudes because seasonal patterns of light and temperature vary dramatically with latitude. In the extreme conditions at the poles, plankton populations crash during the constant darkness of winter and bloom in summer with long hours of light and the retreat of the ice field. In tropical waters, variation in sunlight and temperature is slight, nutrients are present in low concentrations, and planktonic assemblages do not undergo large fluctuations in abundance. There are, however, rapid cycles of reproduction and high rates of grazing and predation that result in a rapid turnover of plankton and a low standing crop. In temperate regions plankton abundance peaks in spring as temperature and the length and intensity of daylight increase. Moreover, seasonal winter storms usually mix the water column, creating a more even distribution of the nutrients, which facilitates the growth of phytoplankton. Peak zooplankton production generally lags behind that of phytoplankton, while the consumption of phytoplankton by zooplankton and phagotrophic protists is thought to reduce phytoplankton abundance. Secondary peaks in abundance occur in autumn. Seasonal peaks of some plankton are very conspicuous, and the composition of the plankton varies considerably. In spring and early summer many fish and invertebrates spawn and release eggs and larvae into the plankton, and, as a result, the meroplanktonic component of the plankton is higher at these times. General patterns of plankton abundance may be further influenced by local conditions. Heavy rainfall in coastal regions (especially areas in which monsoons prevail) can result in nutrient-rich turbid plumes (i.e., estuarine or riverine plumes) that extend into waters of the continental shelf. Changes in production, therefore, may depend on the season, the proximity to fresh water, and the timing and location of upwelling, currents, and patterns of reproduction.

Freshwater Ecosystems

Fresh water is best defined, in contrast to the oceans, as water that contains a relatively small amount of dissolved chemical compounds. Some studies of fresh-water ecosystems focus on water bodies themselves, while others include the surrounding land that interacts with a lake or stream.

The freshwater ecosystems are generally classified into two major groups as, lentic and lotic ecosystems. The term Lentic ecosystem is given to standing water bodies or still water bodies. The LENTIC Ecosystems includes all standing water bodies like Lakes, ponds, swamps or bogs. The term lotic ecosystem is given to the flowing water bodies. The LOTIC Ecosystems include all flowing water bodies like river, springs creek. The subject of study of freshwater ecosystems is known as limnology. Almost all ecological factors like temperature, light, pH, dissolved gases, dissolved salts in water, turbidity, alkalinity, depth and areal distribution all of these parameters play an active role in controlling the habitat of aquatic ecosystems.

1. River Ecosystem
2. Ecological factors of Rivers
3. Life along rivers
4. Lake Ecosystem
5. Ecological factors of Lakes
6. Life in lakes.

River Ecosystem

Water is an essential component of life. Surface water resources are the mostly preferred locations for life settlements. Most of the human civilizations were originated near water courses, especially along the major rivers. A river is a large natural course of flowing water obtained from precipitation. The surface water moves down along the slopes due to the action of gravity.

Streams, tributaries, brooks, creeks and springs are the different types of water courses classified based on their dimension and distribution. A river water is always on the move. Every river has its own longitudinal profile and different cross-sections. The longitudinal profile indicates the nature of slope existing at different places and levels.

The cross-section of a river varies from headwater zone to the mouth. These are called as river valleys which may be ranging from sharp canyons and gorges to wider flat streams nearer to the delta. The velocity of water flowing in a stream is not uniform along the longitudinal profile, also within their cross sections.

A river is a powerful geological agent. It has the capacity to erode, transport and deposit the sediments. These are called as river alluvium. The alluvial deposits, clay and silt of a river are the materials preferred for different activities. A river may be called as major, medium and minor river based on its catchment area, number and length of streams and tributaries, stage of development, and its discharge of water. A river may be called as a perennial river when there is continuous flow of water throughout the year, an intermittent river when the flow is seasonal, an ephemeral river when the flow is occasional or rare.

The following are the terms used to denote the small portions of rivers:

- Pool is a segment where the water is deeper and moving slowly.

- Riffle - is a segment where the flow is shallower and more turbulent.

- Headwater, in a river, is the point of origin of the stream.

- Channel is the river courses developed by constant erosion.

- Floodplain is the flatland existing on either side of the stream that are subject to seasonal flooding.

- The confluence of a river is called as the mouth. This is the point at which the stream discharges all its load into a sea or other static body of water.

A flowing river water carries enormous amount of salts in solution and sediments in suspension. It also rolls up a lot of bed load along the bottom. The water flowing through a river is called as its discharge. The volume and velocity of river discharge depends on several geomorphic factors.

The suspended and bed load sediments carried along with other organic matter in the flowing water control the characteristics of the river ecology. The life along rivers, vary from its head/source to the mouth, from stream to stream, from country to country.

The velocity of flow and force, nature of substratum like alluvium or rock bottom may determine some of the habitat of a river course.

Ecological Factors of Rivers

The ecology of running waters is unique from that of other aquatic habitats. The following are the unique features of Freshwater Aquatic ecosystems:

1. Flow is unidirectional, in lotic ecosystems.

2. There is a state of continuous physical change.

3. There is a high degree of spatial and temporal heterogeneity at all scales (microhabitats).

4. Variability between lotic systems is quite high.

5. The biota is specialized to live with flow conditions.

The major abiotic factors controlling the lotic ecosystems are:

1. Slope and geomorphic conditions including the nature of substratum.

2. Physico-chemical properties of water. Temperature, color, alkalinity, pH and dissolved oxygen.

3. Flow velocity and quantity.

4. Type and amount of suspended and bed-load sediments.

5. Turbidity.

6. Thickness of water column and the depth of light penetration.

7. The climatological factors like atmospheric temperature, humidity, sun shine hours, evapotranspiration and wind.

Depending upon the temperature of water, streams are classified into iso-thermal and non-isothermal streams. In all the rivers, most of the abiotic parameters vary both in space and time. The Interface between the land and water and the interface between water and air play a significant role in controlling the environmental conditions of an area.

Characteristics of Lotic Adaptations

The animals and plants living in lotic environments have certain specific adaptations.

They are subjected to varieties of dynamic environmental factors, like water currents, pollutants and suspended sediments. Lotic habitats are influenced by the effect of continuously moving water, pollution, suspended sediments, floods and other human activities.

The unique characteristics of running water habitat are:

- The establishment of a firm attachment with the substratum. Most of the sponges, diatoms and moss are examples of these. They live on the wooden logs, stones, rock exposures.

- The swimmers are expected to have hooks or suckers to maintain grip over the polished surfaces.

- Some of them build nets around them for food trapping.

- Some of them, like snails and worms, may have sticky bottoms to move longs the base.

- The life living in rivers, have a stream-lined shape of the body. They may have a body rounded anteriorly and tapering posteriorly. This is for a free-swimming habit against the water currents.

- Some have a flat body to stay within the cracks and crevices of rocks.

- Rheotaxis is a feature seen in rivers. This is the capacity, or mechanism by which fishes and other animals swim against the currents and rapidly flowing water.

- This is the resistance capacity of many lotic forms.

- Clinging habitat is another feature of Life in river ecosystems. Some organisms mostly stay closer and nearer to the hard bodies or materials.

- Some of the life forms in rivers have the characteristic feature of Osmoregulation. Especially, the Protozoans eliminate excess water through a contractile vacuole.

- For respiration, life systems in rivers have respiratory siphons. Example: The Mayfly is equipped with gills.

- The productivity is more in streams than standing waters.

- The temperature is not constant along the river course.

- Oxygen content is high at all levels, due to the flowing water.

- CO_2 occurs as carbonate and bicarbonate salts.

- Turbidity is a limiting factor of river ecosystems.

- The pelagic adaptations include both planktonic and nektonic adaptations.

- Floating and swimming organism come under these groups. Planktons possess typical body structures. They are bladder like, needle-like and hair-like.

- Walled bodies and locomotory structures like cilia, appendages, fins and musculature are common to these life.

- Some of them are characterised by light and thin skeletons.

Life Along Rivers

In rivers, there are varieties of life like fishes, plants, animals, and numerous microorganisms that survive. Many of them we may not be able to see. In addition to these, along the river banks, trees and shrubs grow as shelter belts for birds and mammals.

Many tiny organisms exist in river waters. They play a crucial role in maintaining the food supply for the entire ecosystem. They act as feeders, collectors, and grazers. They help in breaking down the plant matter that grows along streams or falling from the overhanging vegetation. The river snails work for processing the calcium present in water to build their shells.

Some of the trees and plants act as shades for other life and filter the pollutants and extract trace metals from the sediments.

- Predator -prey relationships are more along the rivers. The larger fish eats the smaller ones and smaller predatory organisms parasitize the larger fish commonly in rivers.

- Varieties of local and migratory birds, snakes, frogs, bears and other land animals, including cattle and humans, all come to the river for drinking water, fihsing, preparing food, bathing, washing and living.

- Every life along rivers produces waste which becomes food for some other type of feeder.

- The producers or autotrophs are the green plants including the chemosynthetic microorganisms present in rivers.

- The micro consumers of rivers are the herbivores, predators and parasites.

- The decomposers or micro consumers are the worms, bacteria and fungi.

- In a stream ecosystem, food is constantly being produced, consumed and recycled.

- Pollution and other human activities can change the food source and impair the life cycles of the creatures living in and around the water courses.

As all living beings along the river depend on one another, any change in the system parameters will affect all others as well.

Example:

- Floods in rivers.

- Dumping solid wastes into rivers may hamper the normal living environments.

Longitudinal Zonation in Rivers

Streams exhibit two habitats - rapids and pools. So, the stream organisms may be divided into rapid communities and pool communities.

The nature of communities existing in rivers depends on the:

- Type of stream bottom.

- Density of population.

The river bottom may be containing sand, pebbles, clay, bedrock or rubble rock. The rapids community are called as Torrential fauna, as they are subjected to the turbulence created by the currents. Example: A blackfly larva which exists in the rock bottom is an example to this group.

The pools community includes the burrowing types, which are living along the stream banks or bottom. Example: Mayfly nymph and the Dragon-flynymph. There is also a zonation in the stream communities.

The Headwater species are different from the deltaic species. The gradational changes in communities are due to the changes in temperature, velocity of water flow and the quality of water including its pH.

Lake Ecosystem

- A Lake is a large standing water body, surrounded by land.

- The formation of lakes, their physico-chemical conditions and the organisms inhabiting within them, are studied under the branch of science called LIMNOLOGY. In Greek "Limne" means lake or marsh.

- Lakes ecosystem is also called as "lacustrine environment".

- Lakes and reservoirs are more or less closed but mostly dynamic ecosystems.

- Lake, as an ecosystem has several budgets as heat budget, water budget and biomass budget. There are about 1350 lakes and reservoirs in the world.

- Lake water and life are subjected to several natural and man-made threats. Lakes involve complex of interrelated mechanical (currents, waves and sediment transport), Physical (thermal and ice phenomena), Chemical and biological processes.

- Lakes are always under the direct influence of rainwater, river water, sedimentation, biomass and productivity of organisms.

- Lakes ecosystem maintains a state of equilibrium with reference to these factors, which are seasonally varying.

- An important feature of lakes is the evaporation of water from its surface. Sedimentation is a regular process in lakes.

- The movement of water masses occurs in the form of waves, currents, turbulent mixing and wind tides. Mostly they are caused by the wind. The waves have an abrasive power and the capacity to cut off the earth's materials.

- The sediments deposited within the lakes are called as lacustrine deposits.

- Lakes create little worlds of their own.

- Water-plants of all shapes and sizes live under the surface of lakes. Some of the plants are attached to the lake bottom, and others float free. This vegetation provides food for water creatures such as bugs, snails, and fish.

- Lakes are also the favourite haunts of waterfowl such as ducks, geese, swans, flamingos, egrets, cranes, and others.

- Land animals use lakes for drinking water. They also obtain food from lakes in the form of fish, birds, and plant life.

Ecological Factors of Lakes

The stagnant lake water has certain characteristic features with reference to the abiotic and biotic factors.

The light penetration in lakes depends on the turbidity of water.

- The water temperature varies with reference to space (including depth) and time (including time and seasons).

- Only the water at the top is exposed to the air. This leads to decomposition at the bottom. Hence, the dissolved oxygen is relatively low in lake waters, than river waters. It may also decrease with depth.

- The life in lakes, their adaptations and distribution depends on the gradations of oxygen content, light and temperature.

Considering the depth and area of a lake, four distinct ecological zones can be identified in lakes.

They are:

- The littoral zone is the edge areas of the lake extending from the water surface down to 6 to 10 m. This shallow water zone is inhabited by rooted plants. This zone is further divided into Epilimniotic and hypolimniotic zones.

- Epilimniotic zone extends from the water surface up to 6m and the hypolimniotic zone extends from 6m to 10 m.

- Limnetic zone is the open water zone extending to a depth upto the effective light penetration. This is dominated by plankton.

- Profundal zone includes the bottom and the deep water beyond the depth of light penetration. It contains the heterotrophs. This zone exists only in lakes and is normally absent in ponds.

In these ecosystems, the depth of effective light penetration is called as the compensation level. This is the depth at which photosynthesis just balances respiration. The characteristic life in lakes varies with reference to the depth zones. The life in lakes and their abundance, distribution and diversity are influenced by the stratification and movement of oxygen and nutrients.

The energy source is the sunlight. The light penetration is controlled by the turbidity of water. Based on light penetration, the lake can be divided into two zones as:

- Trophogenic zone: This is almost corresponding to the epilimnion zone, dominated by photosynthesis.

- Tropholytic zone: This is a Lower layer wherein the decomposition is expected to be more active. This zone corresponds to the hypolimnion zone.

Based on ecological factors lakes are classified into the following three types:

- Oligotrophic lakes are of very good depths. The water is transparent and the biotic components are poor. The pH of water is low. Nitrogen is negligible. Organic contents are poor. The deeper layers are rich in biota.

- The Eutrophic lakes are shallow water bodies with rich organic matter and plankton and other biota. Such condition comes when the lake is old.

- Dystrophic lakes may be deep or shallow but rich in humus and poor in carbon-di-oxide content. The faunal growth is also poor in such lakes.

Life in Lakes

The organisms living in water may be classified into the following types:

- The surface living organisms, which are called as Planktons, whose movements are mostly controlled by the currents. Algae, protozoa, rotifers, copepods and cladocera belong to this group.

- Animals living at the bottom of water bodies are called as Benthos. These are further divided according to the mode of feeding into filter feeders and deposit-feeders (or sediment feeders). Midge larvae, clams and other microscopic organisms thrive as benthos.

- Active swimming forms called as Nektons. Fishes, aquatic insects, water beetles, amphibians, turtles, water snakes, tadpoles of frogs and Tilapia live as nektons.

- Organisms (both plants and animals) attached or clinging to stems and leaves of rooted plants or projected surfaces. These are called as Periphytons. Sessile algae, fungi, protozoa, hydra, microcrustacea, rotifera and snails come under this category.

- The organisms which are resting and swimming on the surface of water are called as Neustons. Insects, mosquito larvae, some bacteria and algae come under this group.

In a lake or a pond ecosystem, most of these aquatic habitat exist:

- Lake ecosystems are characterised by three adaptations as floating vegetation, submerged vegetation and animal adaptations.

- The floating plants have poorly developed root systems. To remain afloat, the leaves and stems are filled with air-spaces. The upper surface of the leaves is waxy. The floating leaves are larger and broader.

- The submerged vegetation possess certain modifications for their growth and survival. The leaves lack a cuticle.

- The productivity of a lake or pond ecosystem differs depending upon the depth, nutrients, light penetration, light availability and biota. In moderately shallow lakes and ponds, light penetration is more. Organic matter and nutrients also accumulate in heavy amounts. In such water bodies, productivity is higher.

The Aquatic biodiversity is a primary concept in environmental analysis Aquatic ecosystems also provides a home to many species including the phytoplankton, zooplankton, aquatic plants, insects, fish, birds, mammals, and others. In summary, aquatic biodiversity includes all unique species and habitats, and the interaction between them. They are organized at many levels, from the smallest building blocks of life to complete ecosystems, encompassing communities, populations, species, and genetic levels. It has enormous economic and aesthetic value and is largely responsible for maintaining the overall environment.

Humans have long depended on aquatic resources for food, medicines, and materials as well as for recreational and commercial purposes such as fishing and tourism. Several Factors affect these conditions.

They are overexploitation of species, introduction of exotic species, pollution from urban, industrial, and agricultural activities, as well as the habitat loss and alteration through damming and diversion of water into other places.

All these contribute to the declining levels of aquatic biodiversity, especially the freshwater ecosystems. It is necessary to adopt certain conservation strategies to protect and conserve the aquatic life and to maintain the balance of nature and support the availability of resources for future generations.

Terrestrial Ecosystems

A terrestrial ecosystem is an ecosystem that exists on land, rather than on water. Such ecosystem is a community of organisms existing and living together on the land. We can see this from the etymology of the word terrestrial. 'Terrus' is Latin for land. Of course, water may be present in a terrestrial ecosystem. However, terrestrial ecosystems should primarily be situated on land.

In this way, they can be distinguished from marine or fresh water ecosystems, which are located underneath the water.

Features of Terrestrial Ecosystem

- Terrestrial ecosystems are ecosystems that exist on land.
- Etymologically, the word terrestrial comes from the word for land.
- Terrestrial ecosystems are distinct communities of organisms interacting and living together.
- There are many different types of terrestrial ecosystems.
- Terrestrial ecosystems can be distinguished from marine and fresh water ecosystems, which exist under water rather than on land.

Terrestrial ecosystems are distinguished from aquatic ecosystems by the lower availability of water and the consequent importance of water as a limiting factor. Terrestrial ecosystems are characterized by greater temperature fluctuations on both a diurnal and seasonal basis than occur in aquatic ecosystems in similar climates, because water has a high specific heat, a high heat of vaporization, and a high heat of fusion compared with the atmosphere, all of which tend to ameliorate thermal fluctuations. The availability of light is greater in terrestrial ecosystems than in aquatic ecosystems because the atmosphere is more transparent than water. Gases are more available in terrestrial ecosystems than in aquatic ecosystems. Those gases include carbon dioxide that serves as a substrate for photosynthesis, oxygen that serves as a substrate in aerobic respiration, and nitrogen that serves as a substrate for nitrogen fixation. Terrestrial environments are segmented into a subterranean portion from which most water and ions are obtained, and an atmospheric portion from which gases are obtained and where the physical energy of light is transformed into the organic energy of carbon-carbon bonds through the process of photosynthesis.

Major Types of Terrestrial Ecosystems

Desert Ecosystems

The amount of rainfall is the primary abiotic determining factor of a desert ecosystem. Deserts receive less than 25 centimeters (about 10 inches) of rain per year. Large fluctuations between day and night temperatures characterize a desert's terrestrial environment. The soils contain high mineral content with little organic matter.

The vegetation ranges from non-existent to including large numbers of highly adapted plants. The Sonora Desert ecosystem contains a variety of succulents or cactus as well as trees and shrubs. They have adapted their leaf structures to prevent water loss. For instance, the Creosote shrub has a thick layer covering its leaves to prevent water loss due to transpiration.

One of the most famous desert ecosystems is the Sahara desert, which takes up the entire top area of the African continent. The size is comparable to that of the entire United States and is known as the largest hot desert in the world with temperatures reaching over 122 degrees Fahrenheit.

Forest Ecosystems

About one third of the Earth's land is covered in forest. The primary plant in this ecosystem is trees. Forest ecosystems are subdivided by the type of tree they contain and the amount of precipitation they receive.

Some examples of forests are temperate deciduous, temperate rainforest, tropical rainforest, tropical dry forest and northern coniferous forests. Tropical dry forests have wet and dry seasons, while tropical rain forests have rain year-round. Both of these forests suffer from human pressure, such as trees being cleared to make room for farms. Because of the copious amounts of rain and favorable temperatures, rainforests have high biodiversity.

Taiga Ecosystems

Another type of forest ecosystem is the taiga, also known as northern coniferous forest or boreal forest. It covers a large range of land stretching around the northern hemisphere. It is lacking in biodiversity, having only a few species. Taiga ecosystems are characterized by short growing seasons, cold temperatures, and poor soil.

This terrestrial environment has long summer days and very short winter days. Animals found in the taiga include lynx, moose, wolves, bears and burrowing rodents.

Grassland Ecosystems

Temperate grasslands include prairies and steppes. They have seasonal changes, but don't get enough rainfall to support large forests.

Savannas are tropical grasslands. Savannas have seasonal precipitation differences, but temperatures remain constant. Grasslands around the world have been converted to farms, decreasing the amount of biodiversity in these areas. The prominent animals in grassland ecosystems are grazers such as gazelle and antelope.

Tundra

Two types of tundra exist: arctic and alpine. The Arctic tundra is located in the Arctic Circle, north of the boreal forests. Alpine tundras occur on mountain tops. Both types experience cold temperatures throughout the year.

Because the temperatures are so cold, only the top layer of soil in this terrestrial environment thaws during the summer; the rest of it remains frozen year round, a condition known as permafrost. Plants in the tundra are primarily lichens, shrubs, and brush. Tundras do not have trees. Most animals that live in the tundra migrate south or down the mountain for the winter.

Flow of Energy in Ecosystems

Energy has been defined as the capacity to do work. Energy exists in two forms potential and kinetic. Potential energy is the energy at rest (i.e., stored energy) capable of performing work. Kinetic energy is the energy of motion (free energy).

It results in work performance at the expense of potential energy. Conversion of potential energy into kinetic energy involves the imparting of motion. The source of energy required by all living organisms is the chemical energy of their food. The chemical energy is obtained by the conversion of the radiant energy of sun.

The radiant energy is in the form of electromagnetic waves which are released from the sun during the transmutation of hydrogen to helium. The chemical energy stored in the food of living organisms is converted into potential energy by the arrangement of the constituent atoms of food in a particular manner. In any ecosystem there should be unidirectional flow of energy.

Energy Flow in Ecosystems

Living organisms can use energy in two forms radiant and fixed energy. Radiant energy is in the form of electromagnetic waves, such as light. Fixed energy is potential chemical energy bound in various organic substances which can be broken down in order to release their energy content.

Organisms that can fix radiant energy utilizing inorganic substances to produce organic molecules are called autotrophs. Organisms that cannot obtain energy from abiotic source but depend on energy-rich organic molecules synthesized by autotrophs are called heterotrophs. Those which obtain energy from living organisms are called consumers and those which obtain energy from dead organisms are called decomposers.

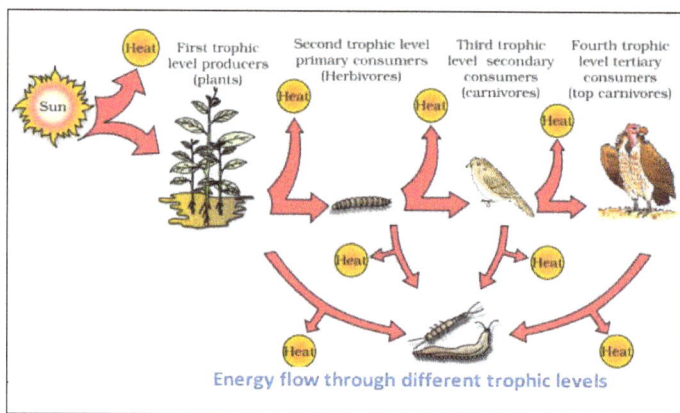

Figure: Flow of energy at different levels of ecosystem.

When the light energy falls on the green surfaces of plants, a part of it is transformed into chemical energy which is stored in various organic products in the plants. When the herbivores consume plants as food and convert chemical energy accumulated in plant products into kinetic energy, degradation of energy will occur through its conversion into heat. When herbivores are consumed by carnivores of the first order (secondary consumers) further degradation will occur. Similarly, when primary carnivores are consumed by top carnivores, again energy will be degraded.

Trophic Level

The producers and consumers in ecosystem can be arranged into several feeding groups, each known as trophic level (feeding level). In any ecosystem, producers represent the first trophic level, herbivores present the second trophic level, primary carnivores represent the third trophic level and top carnivores represent the last level.

Food Chain

In the ecosystem, green plants alone are able to trap in solar energy and convert it into chemical energy. The chemical energy is locked up in the various organic compounds, such as carbohydrates, fats and proteins, present in the green plants. Since virtually all other living organisms depend upon green plants for their energy, the efficiency of plants in any given area in capturing solar energy sets the upper limit to long-term energy flow and biological activity in the community.

The food manufactured by the green plants is utilized by themselves and also by herbivores. Animals feed repeatedly. Herbivores fall prey to some carnivorous animals. In this way one form of life supports the other form. Thus, food from one trophic level reaches to the other trophic level and in this way a chain is established. This is known as food chain.

A food chain may be defined as the transfer of energy and nutrients through a succession of organisms through repeated process of eating and being eaten. In food chain initial link is a green plant or producer which produces chemical energy available to consumers. For example, marsh grass is consumed by grasshopper, the grasshopper is consumed by a bird and that bird is consumed by hawk.

Thus, a food chain is formed which can be written as follows:

Marsh grass → grasshopper → bird → hawk

Food chain in any ecosystem runs directly in which green plants are eaten by herbivores, herbivores are eaten by carnivores and carnivores are eaten by top carnivores. Man forms the terrestrial links of many food chains.

Food chains are of three types:

1. Grazing food chain,

2. Parasitic food chain,

3. Saprophytic or detritus food chain.

Grazing Food Chain

The grazing food chain starts from green plants and from autotrophs it goes to herbivores (primary consumers) to primary carnivores (secondary consumers) and then to secondary carnivores (tertiary consumers) and so on. The gross production of a green plant in an ecosystem may meet three fates—it may be oxidized in respiration, it may be eaten by herbivorous animals and after the death and decay of producers it may be utilized by decomposers and converters and finally released into the environment. In herbivores the assimilated food can be stored as carbohydrates, proteins and fats, and transformed into much more complex organic molecules.

The energy for these transformations is supplied through respiration. As in autotrophs, the energy in herbivores also meets three routes respiration, decay of organic matter by microbes and consumption by the carnivores. Likewise, when the secondary carnivores or tertiary consumers eat primary carnivores, the total energy assimilated by primary carnivores or gross tertiary production follows the same course and its disposition into respiration, decay and further consumption by other carnivores is entirely similar to that of herbivores.

Thus, it is obvious that much of the energy flow in the grazing food chain can be described in terms of trophic levels as outlined below:

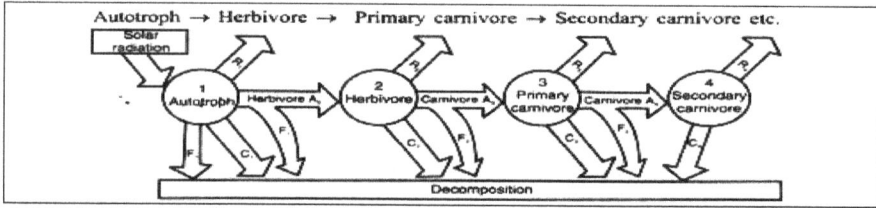

The above figure shows the Diagrammatic representation of a grazing food chain showing input and losses of energy at each trophic level. Trophic levels are numbered and used as subscripts to letters indicating energy transfer. A–assimilation of food by the organisms at the trophic level; F energy lost in the form of faeces and other excretory products: C–energy lost through decay; and R–energy lost to respiration. A schematic representation of grazing food chain showing input and losses of energy has been presented in figure.

Parasitic Food Chain

It goes from large organisms to smaller ones without outright killing as in the case of predator.

Detritus Food Chain

The dead organic remains including metabolic wastes and exudates derived from grazing food chain are generally termed detritus. The energy contained in detritus is not lost in ecosystem as a whole rather it serves as a source of energy for a group of organisms called detritivores that are separate from the grazing food chain. The food chain so formed is called detritus food chain.

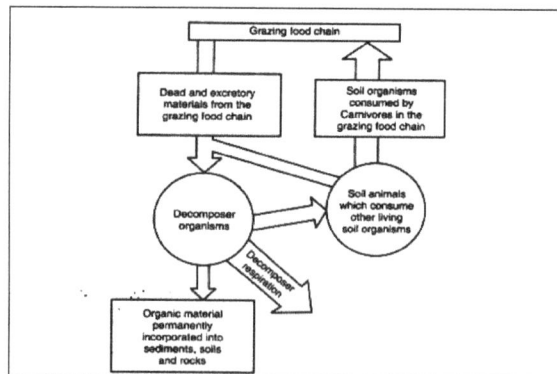

Diagrammatic representation of the detritus food chain showing energy transfers
between it and the grazing food chain, as well as energy losses to the detritus food chain.

In some ecosystems more energy flows through the detritus food chain than through grazing food chain. In detritus food chain the energy flow remains as a continuous passage rather than as a stepwise flow between discrete entities. The organisms in the detritus food chain are many and include algae, fungi, bacteria, slime moulds, actino- mycetes, protozoa, etc. Detritus organisms ingest pieces of partially decomposed or- ganic matter, digest them partially and after extracting some of the chemical energy in the food to run their metabolism, excrete the remainder in the form of simpler organic molecules.

The waste from one organism can be immediately utilized by a second one which re- peats the process. Gradually, the complex organic molecules present in the organic wastes or dead tissues are broken down to much simpler compounds, sometimes to carbon dioxide and water and all that are left are humus. In a normal environment the humus is quite stable and forms an essential part of the soil. Schematic representation of detritus food chain is given in figure.

Food Web

Many food chains exist in an ecosystem, but as a matter of fact these food chains are not independent. In ecosystem, one organism does not depend wholly on another. The resources are shared specially at the beginning of the chain. The marsh plants are eaten by variety of insects, birds, mammals and fishes and some of the animals are eaten by several predators.

Similarly, in the food chain grass → mouse → snakes → owls, sometimes mice are not eaten by snakes but directly by owls. This type of interrelationship interlinks the indi- viduals of the whole community. In this way, food chains become interlinked. A com- plex of interrelated food chains makes up a food web. Food web maintains the stability of the ecosystem. The greater the number of alternative pathways the more stable is the community of living things. Figure illustrates a food web in ecosystem.

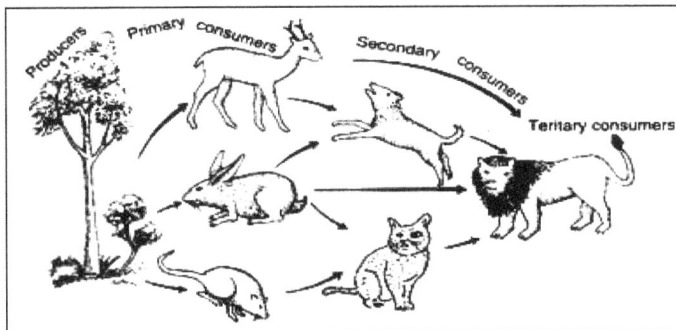

Food web in an ecosystem.

Ecological Pyramid

The trophic structure of an ecosystem can be indicated by means of ecological pyramid. At each step in the food chain a considerable fraction of the potential energy is lost as heat. As a result, organisms in each trophic level pass on lesser energy to the next trophic level than they actually receive. This limits the number of steps in any food chain to 4 or 5. Longer the food chain the lesser energy is available for final members. Because of this tapering off of available energy in the food chain a pyramid is formed that is known as ecological pyramid. The higher the steps in the ecological pyramid the lower will be the number of individuals and the larger their size.

The idea of ecological pyramids was advanced by C.E. Elton. There are different types of ecological pyramids. In each ecological pyramid, producer level forms the base and successive levels make up the apex. Three types of pyramidal relations may be found among the organisms at different levels in the ecosystem.

These are as follows:

1. Pyramid of numbers,

2. Pyramid of biomass (biomass is the weight of living organisms),

3. Pyramid of energy.

Pyramid of Numbers

It depicts the numbers of individuals in producers and in different orders of consumers in an ecosystem. The base of pyramid is represented by producers which are the most abundant. In the successive levels of consumers, the number of organisms goes on decreasing rapidly until there are a few carnivores.

The pyramid of numbers of an ecosystem indicates that the producers are ingested in large numbers by smaller numbers of primary consumers. These primary consumers are eaten by relatively smaller number of secondary consumers and these secondary consumers, in turn, are consumed by only a few tertiary consumers.

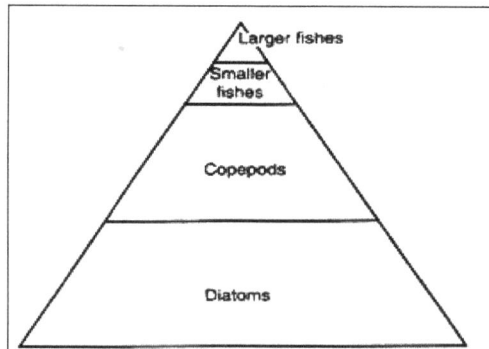

A pyramid of number of a lake ecosystem

This type of pyramid is best presented by taking an example of Lake Ecosystem. In this type of pyramid the base trophic level is occupied by producer elements—algae, diatoms and other hydrophytes which are most abundant. At the second trophic level come the herbivores or zooplanktons which are lesser in number than producers.

The third trophic level is occupied by carnivores which are still smaller in number than the herbivores and the top is occupied by a few top carnivores. Thus, in the ecological pyramid of numbers there is a relative reduction in number of organisms and an increase in the size of body from base to apex of the pyramid. In parasitic food chain starting from tree, the pyramid of numbers will be inverted.

(a & b) Pyramids of number (a) Figure: a & b. Up-right Pyramids of number in a grassland and cultivated field.(b) Pyramid of number (inverted) of diseased tree (Parasitic ecosystem).

Pyramid of Biomass of Organisms

The living weights or biomass of the members of the food chain present at any one time form the pyramid of biomass of organisms. This indicates, by weight or other means of measuring materials, the total bulk of organisms or fixed energy present at one time. Pyramid of biomass indicates the decrease of biomass in each tropic level from base to apex, e.g., total biomass of producers is more than the total biomass of the herbivores.

Likewise, the total biomass of secondary consumers will be lesser than that of herbivores and so on. Since some energy and material are lost in each successive link, the total mass supported at each level is limited by the rate at which the energy is being stored below. This usually gives sloping pyramid for most of the communities in terrestrial and shallow water ecosystems. The pyramid of biomass in a pond ecosystem will be inverted as shown in figure.

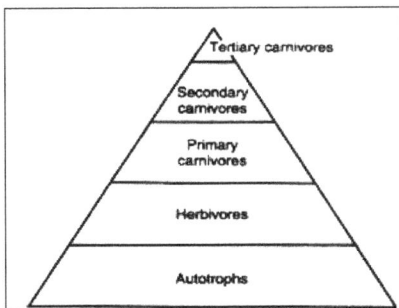

A pyramid of biomass.

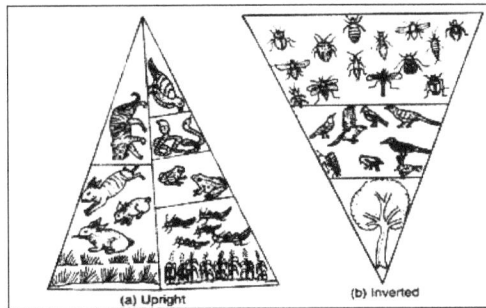

(a & b). Pyramids of biomass (a) A grassland ecosystems showing upright-triangular.
(b) Inverted pyramid of biomass of an aquatic ecosystem.

Pyramid of Energy

This depicts not only the amount of total energy utilized by the organisms at each trophic level of food chain but more important, the actual role of various organisms in transfer of energy. At the producer level the total energy will be much greater than the energy at the successive higher trophic level.

Some producer organisms may have small biomass but the total energy they assimilate and pass on to consumers may be greater than that of organisms with much larger biomass. Higher trophic levels are more efficient in energy utilization but much heat is lost in energy transfer. Energy loss by respiration also progressively increases from lower to higher trophic states.

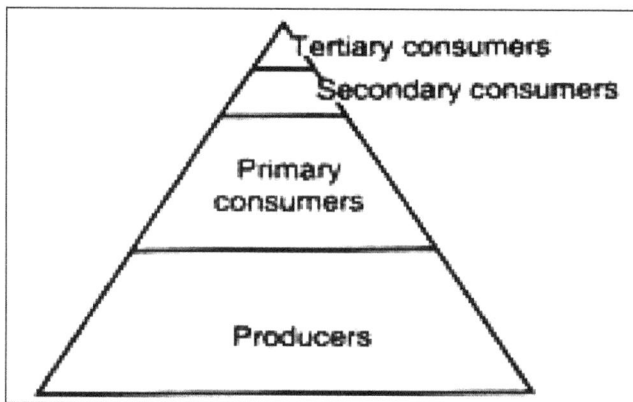

A pyramid of energy.

In the energy flow process, two things become obvious. Firstly there is one way along which energy moves i.e. unidirectional flow of energy. Energy comes in the ecosystem from outside source i.e. sun. The energy captured by autotrophs does not go back to the sun, the energy that passes from autotrophs to herbivores does not revert back and as it moves progressively through the various trophic levels, it is no longer available to the previous levels.

Thus due to unidirectional flow of energy, the system would collapse if the supply from primary source, the sun is cut off. Secondly, there occurs a progressive decrease in energy level at each trophic level which is accounted largely by the energy dissipated as heat in metabolic activities.

Productivity

The relationship between the amount of energy accumulated and the amount of energy utilized within one trophic level of food chain has an important bearing on how much energy from one trophic level passes on to the next trophic level in the food chain. The ratio of output of energy to input of energy is referred to as ecological efficiency.

Different kinds of efficiencies can be measured by the following parameters:

1. Ingestion which indicates the quantity of food or energy taken by trophic level. This is also called exploitation efficiency.

2. Assimilation indicates the amount of food absorbed and fixed into energy rich organic substances which are stored or combined with other molecules to build complex molecules such as proteins, fats etc.

3. Respiration which indicates the energy lost in metabolism.

Primary Productivity

The fraction of fixed energy a trophic level passes on to the next trophic level is called production. Green plants fix solar energy and accumulate it in organic forms as chemical energy. Since it is the first and basic form of energy storage, the rate at which the energy accumulates in the green plants or producers is known as primary productivity.

Primary productivity is the rate at which energy is bound or organic material is created by photosynthesis per unit area of earth's surface per unit time. It is most often expressed as energy in calories/cm^2/yr or dry organic matter in g/m^2/yr (g/m^2 x 8.92 = lb/acre). The amount of organic matter present at a given time per unit area is called standing crop or biomass and as such productivity, which is a rate, is quite different from biomass or standing crop.

The standing crop is usually expressed as dry weight in g/m^2 or kg/m^2 or t/ha (metric tons) or 10^{6}g/hectare. Primary productivity is the result of photosynthesis by green plants including algae of different colours. Bacterial photosynthesis or chemosynthesis, although of small significance may also contribute to primary productivity. The total solar energy trapped in the food material by photosynthesis is referred to as gross primary productivity (G.P.P.).

A good fraction of gross primary production is utilized in respiration of green plants. The amount of energy bound in organic matter per unit area and time that is left after respiration in plants is net primary production (N.P.P.) or plant growth. Only the net primary productivity is available for harvest by man and other animals. Net productivity of energy = gross productivity—energy lost in respiration.

Secondary Productivity

The rates at which the heterotrophic organisms resynthesize the energy-yielding substances is termed as secondary productivity. Secondary productivities are the productivities of animals and saprobes in communities. The amount of energy stored in the tissues of consumers or heterotrophs is termed as net secondary production and the total plant material ingested by herbivores is grass secondary production. Total plant material ingested by herbivores minus the materials lost as faeces is equal to Ingested Secondary Production.

Environmental factors affecting the production processes in an ecosystem are as follows:

1. Solar radiation and Temperature.

2. Moisture: Leaf water potential, soil moisture and precipitation fluctuation and transpiration.

3. Mineral nutrition: Uptake of minerals from the soil, rhizosphere effects, fire effects, salinity, heavy metals, nitrogen metabolism.

4. Biotic activities: Grazing, above ground herbivores, below ground herbivores, predators and parasites, diseases of primary producers.

5. Impact of human population. Pollutions of different sorts, ionizing radiations like atomic explosions, etc.

There are three fundamental concepts of productivity:

1. Standing crop,

2. Materials removed,

3. Production rate.

Standing Crop

It is the abundance of the organisms existing in the area at any one time. It may be expressed in terms of number of individuals, as biomass of organisms, as energy content or in some other suitable terms. Measurement of standing crop reveals the concentration of individuals in the various populations of ecosystem.

The Materials Removed

The second concept of productivity is the materials removed from the area per unit time. It includes the yield to man, organisms removed from the ecosystem by migration, and the material withdrawn as organic deposit.

The Production Rate

The third concept of productivity is the production rate. It is the rate at which the growth processes are going forward within the area. The amount of material formed by each link in the food chain per unit of time per unit area or volume is the production rate.

All the three major groups of organisms—producers, consumers and reducers are the functional kingdoms of natural communities. The three represent major directions of evolution and are characterised by different modes of nutrition. Plants feed primarily by photosynthesis, animals feed primarily by ingesting food that is digested and absorbed in the alimentary canal and the saprobes feed by absorption and have need for an extensive surface of absorption. The principal kinds of organisms among saprobes are the unicellular bacteria, yeasts, chytrids or lower fungi and higher fungi with mycelial bodies.

In terrestrial communities as much as 90% of net primary production remains un-harvested and are utilized as dead tissue by saprobes and soil animals. The saprobes have a larger and more essential role than animals in degrading dead organic matter to inorganic forms and in such ecosystems, secondary production by reducers (decomposers) should exceed that by consumers, though the former is even more difficult to measure than the latter.

Biomass of decomposers with their microscopic cells and filaments embedded in food sources is also difficult to measure and that is small in relation to their productivity and significance for the ecosystem. Small masses of reducers degrade and transform larger masses of organic matter to inorganic remnants. In so doing decomposers disperse back to the environment the energy of photosynthesis accumulated in the organic compounds that are decomposed.

Thus they have a major role in the energy flow of ecosystems. A community or ecosystem, like an organism, is an open energy system. The continuous intake of energy in photosynthesis replaces the energy dissipated to environment by respiration and biological activity and the system does not run-down through the loss of free energy to maximum entropy.

If the amount of energy entrapped is greater than the energy dissipated, the pool of biologically useful energy of organic bonds increases. This results in increase of community biomass and consequently the community grows; such is the case in succession. If energy intake is lesser than energy dissipation, the community biomass will decrease and it must, in some sense, retrogress. If energy intake and loss are in balance, the pool of organic energy is in steady state; such is the case in climax communities.

Three aspects of this steady state may be recognized:

1. The steady state of population of climax communities in which equal birth and death rates in population keep the number of individuals relatively constant,

2. The steady state of energy flow,

3. The steady state of the matter of community, where addition of material by photosynthesis and organic synthesis is balanced by loss of material through respiration and decomposition.

Methods of Measuring Primary Production

There are several parameters for measuring primary production and the methods of measuring primary production are based on those parameters.

The methods are discussed here as under:

1. **Harvest method:** It involves removal of vegetation periodically and weighing the material. For measuring above ground production, the above ground plant parts are clipped at ground level, dried to constant weight at 80°C and weighed. The dry weight in g/m^2 /year gives the ground production. Below ground production is estimated by using frequent core sampling technique of Dahlman and Kucera. It is expressed in terms of weight in gm per unit area per year. In terms of energy one gm dry weight of plant material contains 4 to 5 kcal. The limitations of harvest method are as follows:

   - The amount of plant material consumed by herbivores and the food oxidized during respiration process of the plants is not accounted.

   - Root biomass is neglected.

   - Photosynthetic trans located to underground parts of plants are not known.

   In spite of these limitations the method is used all over for measuring net assimilation rate (NAR) and relative growth rate (RGR).

2. **Carbon dioxide assimilation method:** Utilization of CO_2 in photosynthesis or its liberation during respiration is measured by infrared gas analysis or by passing the gas through Baryta water $Ba(OH)_2$ and titrating the same. The CO_2 removed from incoming gas chamber is taken to be synthesized into organic matter by the green plants. Performing the experiment in light and dark chambers the net and gross production can be measured.

   In the lighted chamber photosynthesis and respiration take place simultaneously and the CO_2 coming out from the chamber is the unused gas of the atmosphere plus gas from the respiration of plant parts. In the dark chamber all CO_2 is due to respiration.

   Net production = Gross production—Respiration

3. **Oxygen production method:** In the aquatic vegetation CO_2 gas analysis method is not used but oxygen evolution method is generally used. The light and dark bottle technique is employed for measuring primary production of aquatic plant. In this method two bottles, one transparent and the other opaque are filled with water at a given depth of lake, closed, maintained at that depth for some time and then brought to laboratory for determination of oxygen content in the water. The decrease of oxygen in dark bottle is due to respiratory activity while increase of O_2 in light bottle is due to photosynthesis. The total increase of O_2 in light bottle plus the amount of O_2 decreased in dark bottle express gross productivity (O_2 value multiplied by 0.375 gives an equivalent of carbon assimilation). Recently, oxygen electrodes have been used for estimating oxygen content in water.

4. Chlorophyll method: Gesner pointed out that the amount of chlorophyll/m^2 is almost limited to a narrow range of 0.1 to 3.0 gm regardless of the age of individuals or the species present therein. There is direct correlation between the amount of chlorophyll and dry matter production in different types of communities with varying light conditions.

The relation of total amount of chlorophyll to the photosynthetic rate is referred to as assimilation ratio or rate of production/gm chlorophyll. Total chlorophyll per unit area is greater in land plants as compared to that in aquatic plants. In marine ecosystem the rate of carbon assimilation is 3.7 g/ hr/g of chlorophyll. The relationship between area based chlorophyll and dry matter production in terrestrial ecosystems has been worked out by Japanese ecologists Argua and Monsi.

Other methods - Pandeya, Sharma and several other ecologists have evolved correlation coefficients for evaluating biomass and productivity in forest trees by measuring their diameter at breast height (DBH), height, canopy cover, etc.

Methods of establishing regression are as below:

1. Diameter of trees in sample quadrats is measured at breast height and the height repeated is determined for each tree.

2. Different diameter and height classes are determined for each species.

3. A set of sample trees are cut and subjected to a detailed analysis for dry weight of stems, twigs, leaves and roots.

4. Regression values are computed for the sets of trees belonging to each girth class, relating the biomass of each fraction to the diameter at breast height.

5. The regression values are used to compute the probable biomass and Production each tree in the sample area. These values for each species when pooled give biomass and production rate of trees per unit area in the forest. Age of the trees markedly influences the annual net production.

Flow of Matter in Ecosystems

The flow of matter in an ecosystem is not like energy flow. Matter enters an ecosystem at any level and leaves at any level. Matter cycles freely between trophic levels and between the ecosystem and the physical environment.

Nutrients

Nutrients are ions that are crucial to the growth of living organisms. Nutrients such as nitrogen and phosphorous are important for plant cell growth. Animals use silica and calcium to build shells and skeletons. Cells need nitrates and phosphates to create proteins and other biochemicals. From nutrients, organisms make tissues and complex molecules such as carbohydrates, lipids, proteins, and nucleic acids.

What are the sources of nutrients in an ecosystem? Rocks and minerals break down to release nutrients. Some enter the soil and are taken up by plants. Nutrients can be brought in from other regions, carried by wind or water. When one organism eats another organism, it receives all of its nutrients. Nutrients can also cycle out of an ecosystem. Decaying leaves may be transported out of an ecosystem by a stream. Wind or water carries nutrients out of an ecosystem.

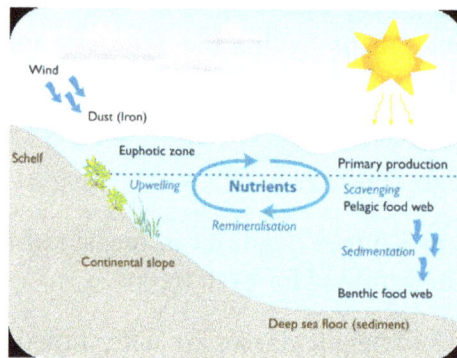

Nutrients cycle through ocean food webs.

Decomposers play a key role in making nutrients available to organisms. Decomposers break down dead organisms into nutrients and carbon dioxide, which they respire into the air. If dead tissue would remain as it is, eventually nutrients would run out. Without decomposers, life on Earth would have died out long ago.

The Food Web

Energy is transferred through an ecosystem in steps, making up a food chain or a food web. At the bottom of the chain are the primary producers, which absorb sunlight and use the light energy to convert carbon dioxide and water into carbohydrates (long chains of sugar molecules) and eventually into other biochemical molecules, by photosynthesis.

The primary producers support the consumers—organisms that ingest other organisms as their food source. Finally, decomposers feed on decaying organic matter, from all levels of the web. Decomposers are largely microscopic organisms (microorganisms) and bacteria.

The food web is really an energy flow system, tracing the path of solar energy through the ecosystem. Solar energy is absorbed by the primary producers and stored in the chemical products of photosynthesis. As these organisms are eaten and digested by consumers, chemical energy is released. This chemical energy is used to power new biochemical re-actions, which again produce stored chemical energy in the consumers' bodies.

Energy is lost at each level in the food web through respiration. You can think of this lost energy as fuel burned to keep the organism operating. Energy expended in respiration is ultimately lost as waste heat and cannot be stored for use by other organisms higher up in the food chain. This means that, generally, both the numbers of organisms and their total amount of living tissue must decrease greatly up the food chain. In general, only 10 to 50 percent of the energy stored in organic matter at one level can be passed up the chain to the next level. Normally, there are about four levels of consumers.

The number of individuals of any species present in an ecosystem depends on the re-sources available to support them. If these resources provide a steady supply of energy, the population size will normally stay steady. But resources can vary with time, for ex-ample, in an annual cycle. In those cases, the population size of a species depending on these resources may fluctuate in a corresponding cycle.

Photosynthesis and Respiration

Simply put, photosynthesis is the production of carbohydrate. Carbohydrate is a gen-eral term for a class of organic compounds that are made from the elements carbon, hydrogen, and oxygen. Carbohydrate molecules are composed of short chains of carbon bonded to one another. Hydrogen (H) atoms and hydroxyl (OH) molecules are also attached to the carbon atoms. We can symbolize a single carbon atom with its attached hydrogen atom and hydroxyl molecule as $-CHOH-$. The leading and trailing dashes indicate that the unit is just one portion of a longer chain of connected carbon atoms. Photosynthesis of carbohydrate requires a series of complex biochemical reactions us-ing water (H_2O) and carbon dioxide (CO_2) as well as light energy. This process requires chlorophyll, a complex organic molecule that absorbs light energy for use by the plant cell. A simplified chemical reaction for photosynthesis can be written as:

$$H_2O + CO_2 + \text{light energy?} -CHOH- + O_2$$

Oxygen gas molecules (O_2) are a by-product of photosynthesis. Because gaseous carbon as CO_2 is "fixed" to a solid form in carbohydrate, we also call photosynthesis a carbon fixation process.

Respiration is the opposite of photosynthesis. In this process, carbohydrate is broken down and combines with oxygen to yield carbon dioxide and water. The overall reac-tion is:

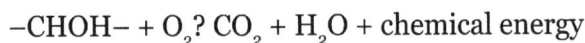

$$-CHOH- + O_2? CO_2 + H_2O + \text{chemical energy}$$

As with photosynthesis, the actual reactions involved are not this simple. The chemical energy released is stored in several types of energy-carrying molecules in living cells and used later to synthesize all the biological molecules used to sustain life.

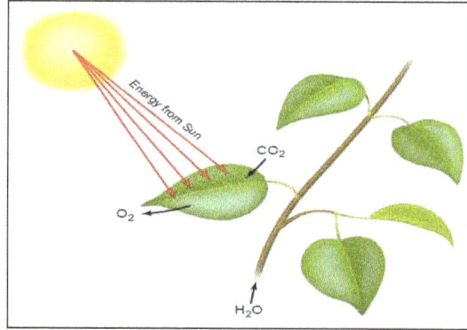
Photosynthesis

Leaves take in CO_2, from the air and H_2O from their roots, using solar energy absorbed by chlorophyll to combine them, forming carbohydrate. In the process, O_2 is released. Photosynthesis takes place in chloroplasts—tiny grains in plant cells that have layers of chlorophyll, enzymes, and other molecules in close contact.

We have to take respiration into account when talking about the amount of new carbohydrate placed in storage. Gross photosynthesis is the total amount of carbohydrate produced by photosynthesis. Net photosynthesis is the amount of synthesized carbohydrate remaining after respiration has broken down sufficient carbohydrate to power the plant:

Net photosynthesis = Gross photosynthesis? Respiration

The rate of net photosynthesis depends on the intensity of light energy available, up to a limit. Most green plants only need about 10 to 30 percent of full summer sunlight for maximum net photosynthesis. Once the intensity of light is high enough for maximum net photosynthesis, the duration of daylight becomes an important factor in determining the rate at which the products of photosynthesis build up in plant tissues. The rate of photosynthesis also increases as air temperature increases, up to a limit.

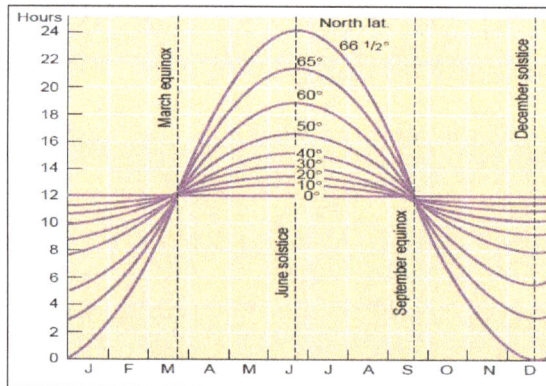
Day-length variation

The graph shows the duration of the daylight period at various latitudes. In the northern hemisphere throughout the year. The angle of the Sun's rays also changes with latitude and the seasons. The vertical scale gives the number of hours the Sun Is above the horizon, with changing seasons. At low latitudes, days are not far from the average 12-hour length throughout the year. At high latitudes, days are short. In winter but long In summer. In subarctic latitudes, photosynthesis can go on in summer during most of the 24-hour day, compensating for the short growing season.

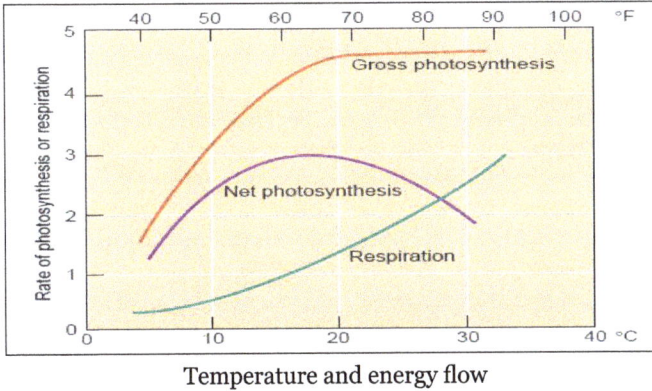

Temperature and energy flow

The above figure shows the results of a laboratory experiment in which sphagnum moss was grown under constant illumination but increasing temperature. Gross photosynthesis increased rapidly to a maximum at about 20 °C (68 °F), then leveled off. But net photosynthesis—the difference between gross photosynthesis and respiration—peaked at about 18 °C (64 °F), then fell off rapidly because respiration continued to increase with temperature.

Net Primary Production

Plant ecologists measure the accumulated net production by photosynthesis in terms of biomass, which is the dry weight of organic matter. This quantity could, of course, be stated for a single plant or animal, but a more useful measurement is the biomass per unit of surface area within the ecosystem—that is, grams of biomass per square meter or (metric) tons of biomass per hectare (1 hectare = 104 m^2). Of all ecosystems, forests have the greatest biomass because of the large amount of wood that the trees accumulate through time. The biomass of grasslands and croplands is much smaller in comparison. The biomass of freshwater bodies and the oceans is about one-hundredth that of the grasslands and croplands. The amount of biomass per unit area tells us about the amount of photosynthetic activity, but it can be misleading.

In some ecosystems, biomass is broken down very quickly by consumers and decomposers. So if we want to know how productive the ecosystem is, it's better to work out the annual yield of useful energy produced by the ecosystem, or the net primary production. Net primary production represents a source of renewable energy derived from the Sun that can be exploited to fill human energy needs. The use of biomass as an energy source involves releasing solar energy that has been fixed in plant tissues through photosynthesis. It can take place in a number of ways—by burning wood for fires, for example.

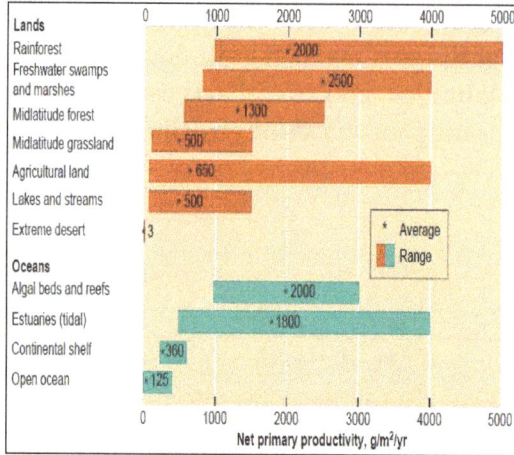

Net primary production of ecosystem: Freshwater swaps and marshes are most productive on land, and algal beds and reefs are most productive in the ocean.

The above figure shows the net primary production of various ecosystems in units of grams of dry organic matter produced annually from one square meter of surface. The highest values are in two quite unlike environments: forests and wetlands (swamps, marshes, and estuaries). Agricultural land compares favorably with grassland, but the range is very large in agricultural land, reflecting many factors such as availability of soil water, soil fertility, and use of fertilizers and machinery.

Open oceans aren't generally very productive. Continental shelf areas are better supporting much of the world's fishing industry. Upwelling zones are also highly productive.

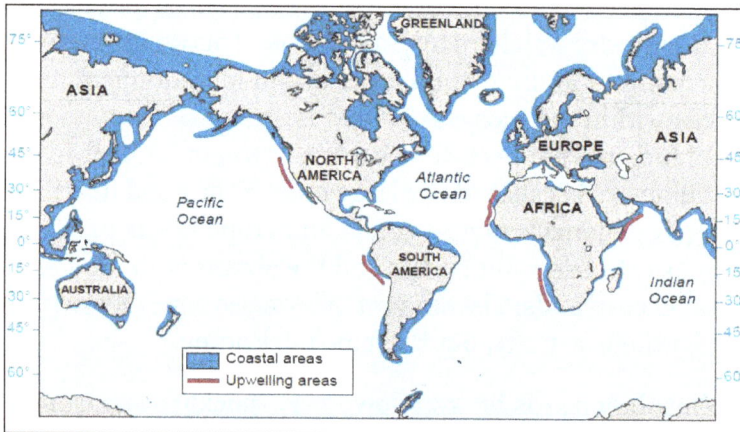

Distribution of world fisheries: Coastal areas and upwelling areas together supply over 99 percent of world production.

The Carbon Cycle

We've seen how energy from the Sun flows through ecosystems, passing from one part of the food chain to the next. Ultimately, that energy is radiated to space and lost from the biosphere. Matter also moves through ecosystems, but because gravity keeps surface material earthbound, matter can't be lost in the global ecosystem. As molecules are formed and re-formed by chemical and biochemical reactions within an ecosystem, the atoms that compose them are not changed or lost. In this way, matter is conserved, and atoms and molecules are used and reused, or cycled, within ecosystems.

Atoms and molecules move through ecosystems under the influence of both physical and biological processes. We call the pathways that a particular type of matter takes through the Earth's ecosystem a biogeochemical cycle (sometimes referred to as a material cycle or nutrient cycle).

The major features of a biogeochemical cycle are diagrammed in figure. Any area or location of concentration of a material is a pool. There are two types of pools: active pools, where materials are in forms and places easily accessible to life processes, and storage pools, where materials are more or less inaccessible to life.

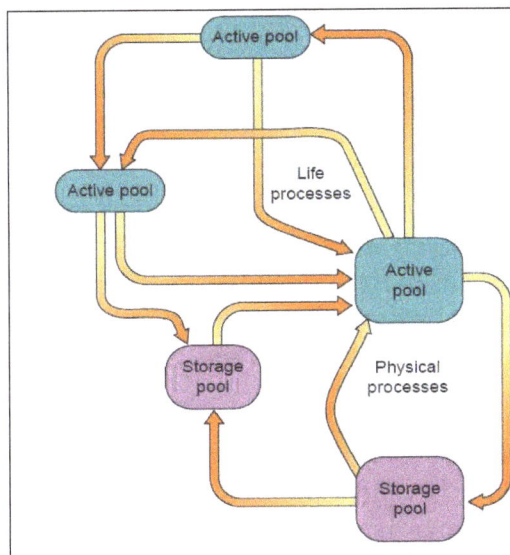

General features of a biogeochemical cycle

A system of pathways of material flows connects the various active and storage pools within the cycle. Pathways can involve the movement of material in all three states of matter—gas, liquid, and solid. For example, carbon moves freely in the atmosphere as carbon dioxide gas and freely in water as dissolved CO_2 and as carbonate ion ($CO_3 =$). It also takes the form of a solid in deposits of limestone and dolomite (calcium and magnesium carbonate).

Ecologists have studied and documented biogeochemical cycles for many elements, including carbon, oxygen, nitrogen, sulfur, and phosphorus. Of these, the carbon cycle is probably the most important. That's because all life is composed of carbon compounds of one form or another and human activities are modifying the carbon cycle in significant ways.

Atmospheric carbon dioxide makes up less than 2 percent of all the carbon. The atmospheric pool is supplied by plant and animal respiration in the oceans and on the lands, by outgassing volcanoes, and by fossil fuel combustion in industry and power generation. The atmosphere loses carbon to oceanic and terrestrial photosynthesis.

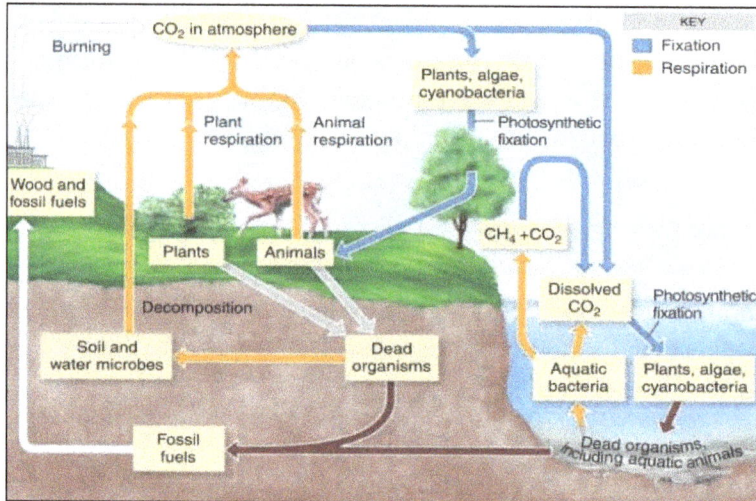

The carbon cycle

Carbon moves through the cycle as a gas a liquid and as a solid in the gaseous portion of the cycle carbon moves largely as carbon dioxide (CO_2) which is a free gas in the atmosphere and a dissolved gas in fresh and saltwater. In the sedimentary portion of its cycle, we find carbon in carbohydrate molecule in organic matter as hydrocarbon compounds in rock (petroleum, coal) and as mineral carbonate compound such as calcium carbonate ($CaCO_2$)

Humans are affecting the carbon cycle by burning fossil fuels. CO_2 is being released to the atmosphere at a rate far beyond that of any natural process, causing global warming. Another important human impact lies in changing the Earth's land covers—for example, in clearing forests or abandoning agricultural areas, or in letting agricultural areas grow back to forests or rangelands.

The Nitrogen Cycle

The nitrogen cycle is another important biogeochemical cycle. Nitrogen makes up 78 percent of the atmosphere by volume, so the atmosphere is a vast storage pool in this cycle. The figure below diagrams the nitrogen cycle. Nitrogen as N_2 in the atmosphere can't be assimilated directly by plants or animals. But certain microorganisms, including some soil bacteria and blue-green algae, can change N_2 into useful forms in a process called nitrogen fixation. Legumes—such as clover, alfalfa, soybeans, peas, beans, and peanuts—are also able to fix nitrogen, with help from bacteria. They have a symbiotic relationship with bacteria of the genus Rhizobium, which is associated with some 190 species of trees and shrubs. The bacteria infect these plants' root cells and supply nitrogen to the plant through nitrogen fixation, while the plants supply nutrients and organic compounds needed by the bacteria. Crops of legumes are often planted in seasonal rotation with other food crops to ensure an adequate nitrogen supply in the soil. Other soil bacteria convert nitrogen from usable forms back to N_2, in a process called denitrification that returns the nitrogen to the atmosphere. Other processes shown in the figure are ammonification, nitrification, and assimilation.

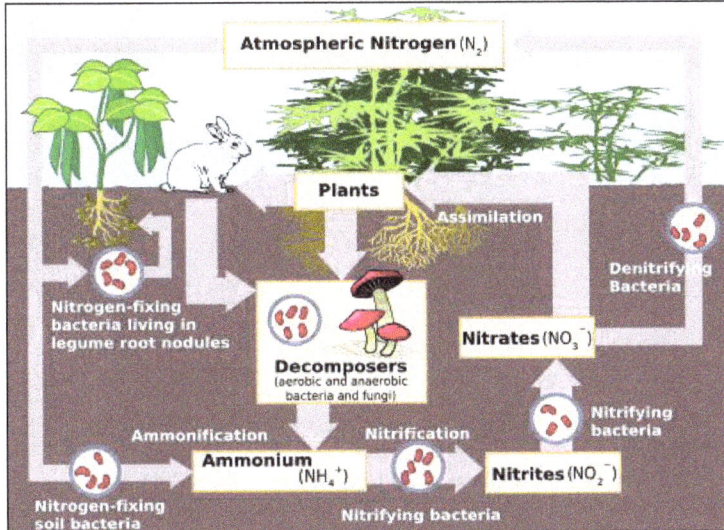

The nitrogen cycle: The five processes of the nitrogen cycle are nitrogen fixation, nitrification, assimilation, ammonification, and denitrification.

At the present time, nitrogen fixation far exceeds denitrification thanks to human activity. We fix nitrogen in the manufacture of nitrogen fertilizers; by oxidizing nitrogen in the combustion of fossil fuels; and through the widespread cultivation of legumes. At present rates, nitrogen fixation from human activity nearly equals all natural biological fixations and usable nitrogen is accumulating in the Earth's ecosystems.

Much of this newly fixed nitrogen is carried from the soil into rivers and lakes and ultimately to the ocean, causing water pollution. The nitrogen stimulates the growth of algae and phytoplankton, which in turn reduce quantities of dissolved oxygen through respiration. Oxygen then drops to levels that are too low for many desirable forms of aquatic life. These problems will be accentuated in years to come because industrial fixation of nitrogen in fertilizer manufacture is doubling about every six years at present. The global impact of such large amounts of nitrogen reaching rivers, lakes, and oceans on the Earth's global ecosystem remains uncertain.

References

- What-is-an-ecosystem: conserve-energy-future.com, Retrieved 10 June, 2019

- Types-aquatic-ecosystems-6123685: sciencing.com, Retrieved 14 February, 2019

- Marine-ecosystem, science: britannica.com, Retrieved 24 April, 2019

- Aquatic-ecosystems-freshwater-types: researchgate.net, Retrieved 2 July, 2019

- Terrestrial-ecosystem-its-meaning-and-types: importantindia.com, Retrieved 12 January, 2019

- Terrestrial-ecosystem: accessscience.com, Retrieved 10 August, 2019

- Major-types-terrestrial-ecosystems-8248888: sciencing.com, Retrieved 15 March, 2019

- Energy-flow-in-an-ecosystem, ecosystem: biologydiscussion.com, Retrieved 29 June, 2019

- Flow-of-Matter-in-Ecosystems, earth-science: ck12.org, Retrieved 9 January, 2019

- Energy-and-matter-flow-in-ecosystems: geography.name, Retrieved 11 May, 2019

Earth's Elements

There are varied elements which are found on the Earth such as minerals, rocks, soil and water. The minerals are classified into numerous categories such as nonmetals, metals and semimetals. This chapter has been carefully written to provide an easy understanding of the properties and types of these elements.

Minerals

Mineral is naturally occurring homogeneous solid with a definite chemical composition and a highly ordered atomic arrangement; it is usually formed by inorganic processes. There are several thousand known mineral species, about 100 of which constitute the major mineral components of rocks; these are the so-called rock-forming minerals.

Nature of Minerals

Morphology

Nearly all minerals have the internal ordered arrangement of atoms and ions that is the defining characteristic of crystalline solids. Under favourable conditions, minerals may grow as well-formed crystals, characterized by their smooth plane surfaces and regular geometric forms. Development of this good external shape is largely a fortuitous outcome of growth and does not affect the basic properties of a crystal. Therefore, the term crystal is most often used by material scientists to refer to any solid with an ordered internal arrangement, without regard to the presence or absence of external faces.

Azurite: Azurite crystals

Symmetry Elements

The external shape, or morphology, of a crystal is perceived as its aesthetic beauty, and its geometry reflects the internal atomic arrangement. The external shape of well-formed crystals expresses the presence or absence of a number of symmetry elements. Such symmetry elements include rotation axes, rotoinversion axes, a centre of symmetry, and mirror planes.

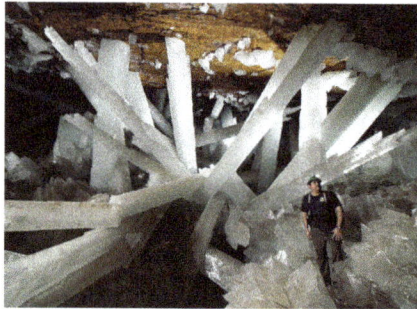

Cueva de los Cristales: Massive selenite (gypsum) crystals from the Cave of Crystals (Cueva de los Cristales), Naica Mine, Chihuahua, Mexico.

A rotation axis is an imaginary line through a crystal around which it may be rotated and repeat itself in appearance one, two, three, four, or six times during a complete rotation. (For example, a sixfold rotation occurs when the crystal repeats itself each 60°—that is, six times in a 360° rotation).

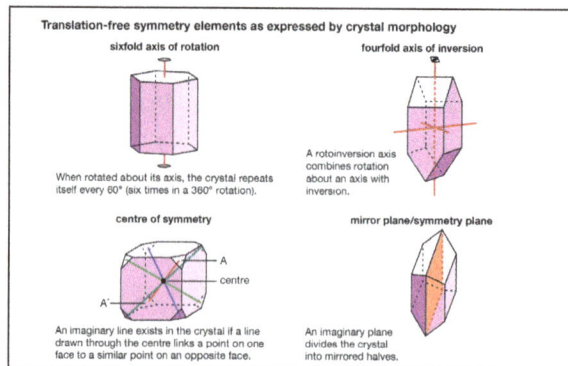

Symmetry elements: Translation-free symmetry elements as expressed by the morphology of crystals.

A rotoinversion axis combines rotation about an axis of rotation with inversion. Rotoinversion axes are symbolized as, $\bar{1}$, $\bar{2}$, $\bar{3}$, $\bar{4}$, and $\bar{6}$, where $\bar{1}$ is equivalent to a centre of symmetry (or inversion), $\bar{2}$ is equivalent to a mirror plane, and $\bar{3}$ is equivalent to a threefold rotation axis plus a centre of symmetry. When the axis of the crystal is vertical, $\bar{4}$ is characterized by two top faces with identical faces upside down underneath. $\bar{6}$ is equivalent to a threefold rotation axis with a mirror plane perpendicular to the axis.

A centre of symmetry exists in a crystal if an imaginary line can be extended from any point on its surface through its centre and a similar point is present along the line equidistant from the centre. This is equivalent to 1, or inversion. There is a relatively simple procedure for recognizing a centre of symmetry in a well-formed crystal. With the crystal laid down on any face on a table top, the presence of a face of equal size and shape, but inverted, in a horizontal position at the top of the crystal proves the existence of a centre of symmetry. An imaginary mirror plane (or symmetry plane) can also be used to separate a crystal into halves. In a perfectly developed crystal, the halves are mirror images of one another.

Morphologically, crystals can be grouped into 32 crystal classes that represent the 32 possible symmetry elements and their combinations. These crystal classes, in turn, are grouped into six crystal systems. In decreasing order of overall symmetry content, beginning with the system with the highest and most complex crystal symmetry, they are isometric (or cubic), hexagonal, tetragonal, orthorhombic, monoclinic, and triclinic. (Many sources list seven crystal systems by dividing the hexagonal crystal system into two parts—trigonal and hexagonal).

The 32 Crystal Classes and their Symmetry Contents		
Crystal System	Symmetry Content	Crystal Class
Triclinic	None	1
	I	1
Monoclinic	$1a_2$	2
	1m	M
	I, $1A_2$, 1m	2/m
Orthorhombic	$3a_2$	222
	A_2, 2m	Mm2
	I, $3A_2$, 3m	2/m2/m2/m
Tetragonal	$1a_4$	4
	$1A_4$	4
	I, $1A_4$, m	4/m
	$1A_4$, $4A_2$	422
	$1A_4$, 4m	4mm
	$1A_4$, $2A_2$, 2m	42m
	I, $1A_4$, $4A_2$, 5m	4/m2/m2/m
Hexagonal	$1a_3$	3
	$1A_3$ $(= i + 1A_3)$	3
	$1A_3$, $3A_2$	32
	$1A_3$, 3m	3m
	$1A_3$, $3A_2$, 3m $(1A_3 = i + 1A_3)$	32/m
	$1A_6$	6
	$1A_6$ $(= 1A_3 + m)$	6

	$I, 1A_6, 1m$	$6/m$
	$1A_6, 6A_2$	622
	$1A_6, 6m$	$6mm$
	$1A_6, 3A_2, 3m \ (1A_6 = 1A_3 + m)$	$6m2$
	$I, 1A_6, 6A_2, 7m$	$6/m2/m2/m$
Isometric	$3a_2, 4a_3$	23
	$3A_2, 3m, 4A_3 \ (1A_3 = 1A_3 + i)$	$2/m3$
	$3A_4, 4A_3, 6A_2$	432
	$3A_4, 4A_3, 6m$	$43m$
	$3A_4, 4A_3, 6A_2, 9m \ (1A_3 = 1A_3 + i)$	$4/m32/m$

The c axis is normally the vertical axis. The isometric system exhibits three mutually perpendicular axes of equal length (a_1, a_2, and a_3). The orthorhombic and tetragonal systems also contain three mutually perpendicular axes; in the former system all the axes are of different lengths (a, b, and c), and in the latter system two axes are of equal length (a_1 and a_2) while the third (vertical) axis is either longer or shorter (c). The hexagonal system contains four axes: three equal-length axes (a_1, a_2, and a_3) intersect one another at 120° and lie in a plane that is perpendicular to the fourth (vertical) axis of a different length. Three axes of different lengths (a, b, and c) are present in both the monoclinic and triclinic systems. In the monoclinic system, two axes intersect one another at an oblique angle and lie in a plane perpendicular to the third axis; in the triclinic system, all axes intersect at oblique angles.

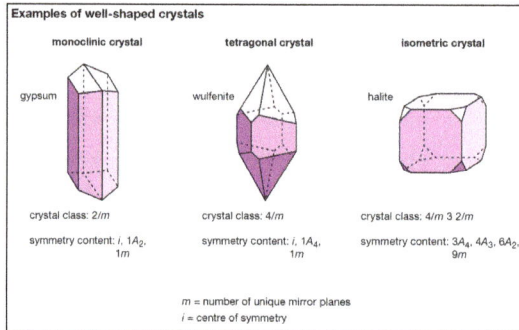

Examples of well-shaped crystals: Each of these examples of well-shaped crystals belongs to a different crystal class because of its overall symmetry content.

Twinning

If two or more crystals form a symmetrical intergrowth, they are referred to as twinned crystals. A new symmetry operation (called a twin element), which is lacking in a single untwinned crystal, relates the individual crystals in a twinned position. There are three twin elements that may relate the crystals of a twin: (1) reflection by a mirror plane (twin plane), (2) rotation about a crystal direction common to both (twin axis) with the angular rotation typically 180°, and (3) inversion about a point (twin centre).

An instance of twinning is defined by a twin law that specifies the presence of a plane, an axis, or a centre of twinning. If a twin has three or more parts, it is referred to as a multiple, or repeated, twin.

Internal Structure

Examining Crystal Structures

The external morphology of a mineral is an expression of the fundamental internal architecture of a crystalline substance—i.e., its crystal structure. The crystal structure is the three-dimensional, regular arrangement of chemical units (atoms, ions, and anionic groups in inorganic materials; molecules in organic substances); these chemical units (referred to here as motifs) are repeated by various translational and symmetry operations. The morphology of crystals can be studied with the unaided eye in large well-developed crystals and has been historically examined in considerable detail by optical measurements of smaller well-formed crystals through the use of optical goniometers (instruments that measure the angles between crystal faces). The internal structure of crystalline materials, however, is revealed by a combination of X-ray, neutron, and electron diffraction techniques, supplemented by a variety of spectroscopic methods, including infrared, optical, Mössbauer, and resonance techniques. These methods, used singly or in combination, provide a quantitative three-dimensional reconstruction of the location of the atoms (or ions), the chemical bond types and their positions, and the overall internal symmetry of the structure. The repeat distances in most inorganic structures and many of the atomic and ionic motif sizes are on the order of 1 to 10 angstroms (Å; 1 Å is equivalent to 10^{-8} cm or 3.94×10^{-9} inch) or 10 to 100 nanometres (nm; 1 nm is equivalent to 10^{-7} cm or 10 Å).

Space Groups

Symmetry elements that are observable in the external morphology of crystals, such as rotation and rotoinversion axes, mirror planes, and a centre of symmetry, also are present in their internal atomic structure. In addition to these symmetry elements, there are translations and symmetry operations combined with translations. (Translation is the operation in which a motif is repeated in a linear pattern at intervals that are equal to the translation distance [commonly on the 1 to 10 Å level].) Two examples of translational symmetry elements are screw axes (combining rotation and translation) and glide planes (combining mirroring and translation). The internal translation distances are exceedingly small and can be seen directly only by very high-magnification electron beam techniques, as used in a transmission electron microscope, at magnifications of about 600,000×.

When all possible combinations of translational elements compatible with the 32 crystal classes (also known as point groups) are considered, one arrives at 230 possible ways in which translations, translational symmetry elements (screw axes and glide planes), and translation-free symmetry elements (rotation and rotoinversion axes and mirror planes) can be combined. These translation and symmetry groupings are known as the 230 space groups, representing the various ways in which motifs can be arranged in an ordered three-dimensional array. The symbolic representation of space groups is closely related to that of the Hermann-Mauguin notation of point groups. A detailed discussion of space groups, their derivation, and notation is beyond the scope of this article. For more specific information, consult the books on mineralogy cited in the Bibliography.

Illustrating Crystal Structures

The external morphology of three-dimensional arrangement of crystal structures may be presented on a two-dimensional page or within a computer simulation. Another common method of illustration involves projecting the crystal structure onto a planar surface. The high-temperature form of silicon dioxide (SiO_2) known as tridymite may be represented this way; however, the structural motif units in this case are SiO_4 tetrahedrons composed of a silicon atom surrounded by four oxygen atoms. To further aid the visualization of complex crystal structures within the physical world, three-dimensional physical models of such structures can be built or obtained commercially. Models of this sort reproduce the internal atomic arrangement on an enormously enlarged scale (e.g., one angstrom might be represented by one centimetre).

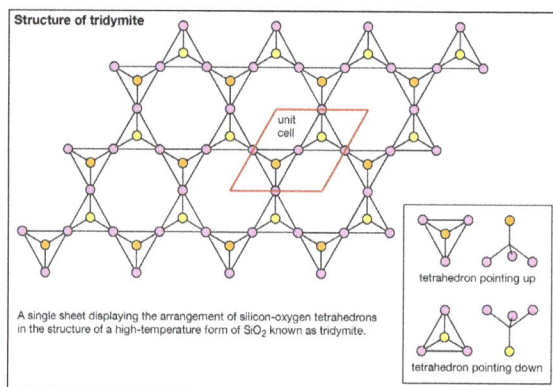

Silicon-oxygen tetrahedrons: Single sheet displaying the arrangement of the silicon-oxygen tetrahedrons in the structure of a high-temperature form of sio2 known as tridymite.

Polymorphism

Polymorphism is the ability of a specific chemical composition to crystallize in more than one form. This generally occurs as a response to changes in temperature or pressure or both. The different structures of such a chemical substance are called polymorphic forms, or polymorphs. For example, the element carbon (C) occurs in nature in two different polymorphic forms, depending on the external (pressure and temperature) conditions. These forms are graphite, with a hexagonal structure, and diamond, with an isometric structure. The composition FeS_2 occurs most commonly as pyrite, with an isometric structure, but it is also found as marcasite, which has an orthorhombic internal arrangement. The composition SiO_2 is found in a large number of polymorphs, among them quartz, tridymite, cristobalite, coesite, and stishovite. The stability field (conditions under which a mineral is stable) of these SiO_2 polymorphs can be expressed in a stability diagram, with the external parameters of temperature and pressure as the two axes. In the general quartz field, there is additional polymorphism leading to the notation of high quartz and low quartz, each form having a slightly different internal structure. Cristobalite and tridymite are the high-temperature forms of SiO_2, and indeed these SiO_2 polymorphs occur in high-temperature lava flows. The high-pressure forms of SiO_2 are coesite and stishovite, and these can be found in meteorite craters, formed as a result of high explosive pressures upon quartz-rich sandstones, and in very deep-seated rock formations, as from Earth's upper mantle or very deep in subduction zones.

Pyrite: Pyrite from Navajun, Spain

Chemical Composition

The chemical composition of a mineral is of fundamental importance because its properties greatly depend on it. Such properties, however, are determined not only by the chemical composition but also by the geometry of the constituent atoms and ions and by the nature of the electrical forces that bind them. Thus, for a complete understanding of minerals, their internal structure, chemistry, and bond types must be considered.

Various analytical techniques may be employed to obtain the chemical composition of a mineral. Quantitative chemical analyses mainly use so-called wet analytical methods (e.g., dissolution in acid, flame tests, and other classic techniques of bench chemistry that rely on observation), in which the mineral sample is first dissolved. Various compounds are then precipitated from the solution, which are weighed to obtain a gravimetric analysis. A number of analytical procedures have been introduced that provide faster but somewhat less accurate results. Most analyses use instrumental methods such as optical emission, X-ray fluorescence, atomic absorption spectroscopy, and electron microprobe analysis. Relatively well-established error ranges have been documented for these methods, and samples must be prepared in a specific manner for each technique. A distinct advantage of wet analytical procedures is that they make it possible to determine quantitatively the oxidation states of positively charged atoms, called cations (e.g., Fe^{2+} versus Fe^{3+}), and to ascertain the amount of water in hydrous minerals. It is more difficult to provide this type of information with instrumental techniques.

To ensure an accurate chemical analysis, the selected sample, which might include several minerals, is often made into a thin section (a section of rock less than 1 mm thick cemented for study between clear glass plates). To reduce the effect of the impurities, an instrumental technique, such as electron microprobe analysis, is commonly employed. In this method, quantitative analysis in situ may be performed on mineral grains only 1 micrometre (10^{-4} centimetre) in diameter.

Mineral Formulas

Elements may exist in the native (uncombined) state, in which case their formulas are simply their chemical symbols: gold (Au), carbon (C) in its polymorphic form of diamond, and sulfur (S) are common examples. Most minerals, however, occur as compounds consisting of two or more elements; their formulas are obtained from quantitative chemical analyses and indicate the relative proportions of the constituent elements. The formula of sphalerite, ZnS, reflects a one-to-one ratio between atoms of zinc and those of sulfur. In bornite (Cu_5FeS_4), there are five atoms of copper (Cu), one atom of iron (Fe), and four atoms of sulfur. There exist relatively few minerals with constant composition; notable examples include quartz (SiO_2) and kyanite (Al_2SiO_5). Minerals of this sort are termed pure substances. Most minerals display considerable variation in the ions that occupy specific atomic sites within their structure. For example, the iron content of rhodochrosite ($MnCO_3$) may vary over a wide range. As ferrous iron (Fe^{2+}) substitutes for manganese cations (Mn^{2+}) in the rhodochrosite structure, the formula for the mineral might be given in more general terms—namely, $(Mn, Fe)CO_3$. The amounts of manganese and iron are variable, but the ratio of the cation to the negatively charged anionic group remains fixed at one Mn^{2+} or Fe^{2+} atom to one CO_3 group.

Sphalerite: Sphalerite, a mineral that is the principal ore of zinc.

Compositional Variation

Most minerals exhibit a considerable range in chemical composition. Such variation results from the replacement of one ion or ionic group by another in a particular structure. This phenomenon is termed ionic substitution, or solid solution. Three types of solid solution are possible, and these may be described in terms of their corresponding mechanisms—namely, substitutional, interstitial, and omission.

Substitutional solid solution is the most common variety. In the carbonate mineral rhodochrosite ($MnCO_3$), Fe^{2+} may substitute for Mn^{2+} in its atomic site in the structure.

The degree of substitution may be influenced by various factors, with the size of the ion being the most important. Ions of two different elements can freely replace one another only if their ionic radii differ by approximately 15 percent or less. Limited substitution can occur if the radii differ by 15 to 30 percent, and a difference of more than 30 percent makes substitution unlikely. These limits, calculated from empirical data, are only approximate.

The temperature at which crystals grow also plays a significant role in determining the extent of ionic substitution. The higher the temperature, the more extensive is the thermal disorder in the crystal structure and the less exacting are the spatial requirements. As a result, ionic substitution that could not have occurred in crystals grown at low temperatures may be present in those grown at higher ones. The high-temperature form of $KAlSi_{3}O_8$ (sanidine), for example, can accommodate more sodium (Na) in place of potassium (K) than can microcline, its low-temperature counterpart.

An additional factor affecting ionic substitution is the maintenance of a balance between the positive and negative charges in the structure. Replacement of a monovalent ion (e.g., Na^+, a sodium cation) by a divalent ion (e.g., Ca^{2+}, a calcium cation) requires further substitutions to keep the structure electrically neutral.

Simple cationic or anionic substitutions are the most basic types of substitutional solid solution. A simple cationic substitution can be represented in a compound of the general form A^+X^- in which cation B^+ replaces in part or in total cation A^+. Both cations in this example have the same valence (+1), as in the substitution of K^+ (potassium ions) for Na^+ (sodium ions) in the NaCl (sodium chloride) structure. Similarly, the substitution of anion X^- by Y^- in an A^+X^- compound represents a simple anionic substitution; this is exemplified by the replacement of Cl^- (chlorine ions) with Br^- (bromine ions) in the structure of KCl (potassium chloride). A complete solid-solution series involves the substitution in one or more atomic sites of one element for another that ranges over all possible compositions and is defined in terms of two end-members. For example, the two end-members of olivine [$(Mg, Fe)_2SiO_4$], forsterite (Mg_2SiO_4) and fayalite (Fe_2SiO_4), define a complete solid-solution series (called the forsterite-fayalite series) in which magnesium cations (Mg^{2+}) are replaced partially or totally by Fe^{2+}.

In some instances, a cation B^{3+} may replace some A^{2+} of compound $A^{2+}X^{2-}$. So that the compound will remain neutral, an equal amount of A^{2+} must concurrently be replaced by a third cation, C^+. This is given in equation form as $2A^{2+} \longleftrightarrow B^{3+} + C^+$; the positive charge on each side is the same. Substitutions such as this are termed coupled substitutions. The plagioclase feldspar series exhibits complete solid solution, in the form of coupled substitutions, between its two end-members, albite ($NaAlSi_3O_8$) and anorthite ($CaAl_2Si_2O_8$). Every atomic substitution of Na^+ by Ca^{2+} is accompanied by the replacement of a silicon cation (Si^{4+}) by an aluminum cation (Al^{3+}), thereby maintaining electrical neutrality: $Na^+ + Si^{4+} \longleftrightarrow Ca^{2+} + Al^{3+}$.

The second major type of ionic substitution is interstitial solid solution, or interstitial substitution. It takes place when atoms, ions, or molecules fill the interstices (voids) found between the atoms, ions, or ionic groups of a crystal structure. The interstices may take the form of channel-like cavities in certain crystals, such as the ring silicate beryl ($Be_3Al_2Si_6O_{18}$). Potassium, rubidium (Rb), cesium (Cs), and water, as well as helium (He), are some of the large ions and gases found in the tubular voids of beryl.

The least common type of solid solution is omission solid solution, in which a crystal contains one or more atomic sites that are not completely filled. The best-known example is exhibited by pyrrhotite ($Fe_{1-x}S$). In this mineral, each iron atom is surrounded by six neighbouring sulfur atoms. If every iron site in pyrrhotite were occupied by ferrous iron, its formula would be FeS. There are, however, varying percentages of vacancy in the iron site, so that the formula is given as Fe_6S_7 through $Fe_{11}S_{12}$, the latter being very near to pure FeS. The formula for pyrrhotite is normally written as $Fe_{1-x}S$, with x ranging from 0 to 0.2. It is one of the minerals referred to as a defect structure, because it has a structural site that is not completely occupied.

Chemical Bonding

Electrical forces are responsible for the chemical bonding of atoms, ions, and ionic groups that constitute crystalline solids. The physical and chemical properties of minerals are attributable for the most part to the types and strengths of these binding forces; hardness, cleavage, fusibility, electrical and thermal conductivity, and the coefficient of thermal expansion are examples of such properties. On the whole, the hardness and melting point of a crystal increase proportionally with the strength of the bond, while its coefficient of thermal expansion decreases. The extremely strong forces that link the carbon atoms of diamond, for instance, are responsible for its distinct hardness. Periclase (MgO) and halite (NaCl) have similar structures; however, periclase has a melting point of 2,800 °C (5,072 °F) whereas halite melts at 801 °C (1,474 °F). This discrepancy reflects the difference in the bond strength of the two minerals: since the atoms of periclase are joined by a stronger electrical force, a greater amount of heat is needed to separate them.

The electrical forces, called chemical bonds, can be divided into five types: ionic, covalent, metallic, van der Waals, and hydrogen bonds. Classification in this manner is largely one of expediency; the chemical bonds in a given mineral may in fact possess characteristics of more than one bond type. For example, the forces that link the silicon and oxygen atoms in quartz exhibit in nearly equal amount the characteristics of both ionic and covalent bonds. As stated above, the electrical interaction between the atoms of a crystal determines its physical and chemical properties. Thus, classifying minerals according to their electrical forces will cause those species with similar properties to be grouped together. This fact justifies classification by bond type.

Ionic Bonds

Atoms have a tendency to gain or lose electrons so that their outer orbitals become stable; this is normally accomplished by these orbitals being filled with the maximum allowed number of valence electrons. Metallic sodium, for example, has one valence electron in its outer orbital; it becomes ionized by readily losing this electron and exists as the cation Na^+. Conversely, chlorine gains an electron to complete its outer orbital, thereby forming the anion Cl^-. In the mineral halite, NaCl (common, or rock, salt), the chemical bonding that holds the Na^+ and Cl^- ions together is the attraction between the two opposite charges. This bonding mechanism is referred to as ionic, or electrovalent.

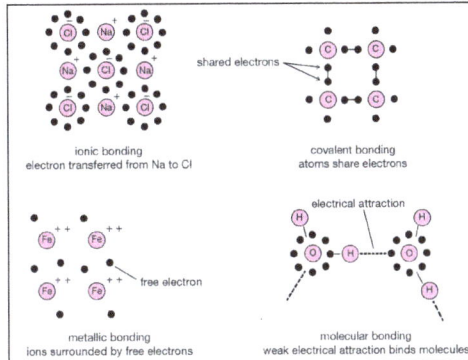

Crystal bonding: Different types of bonding in crystals

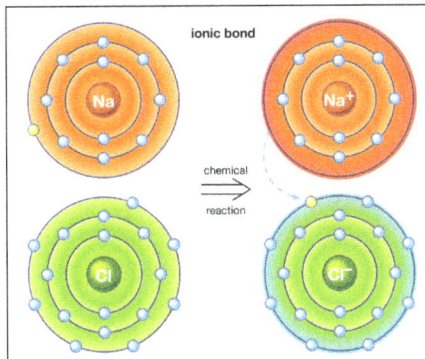

Ionic bond: sodium chloride, or table salt: Ionic bonding in sodium chloride. An atom of sodium (na) donates one of its electrons to an atom of chlorine (cl) in a chemical reaction, and the resulting positive ion (na+) and negative ion (cl−) form a stable ionic compound (sodium chloride; common table salt) based on this ionic bond.

Ionically bonded crystals typically display moderate hardness and specific gravity, rather high melting points, and poor thermal and electrical conductivity. The electrostatic charge of an ion is evenly distributed over its surface, and so a cation tends to become surrounded with the maximum number of anions that can be arranged around it. Since ionic bonding is non-directional, crystals bonded in this manner normally display high symmetry.

Covalent Bonds

In the discussion of the ionic bond, it was noted that chlorine readily gains an electron to achieve a stable electron configuration. An incomplete outer orbital places a chlorine atom in a highly reactive state, so it attempts to combine with nearly any atom in its proximity. Because its closest neighbour is usually another chlorine atom, the two may bond together by sharing one pair of electrons. As a result of this extremely strong bond, each chlorine atom enters a stable state.

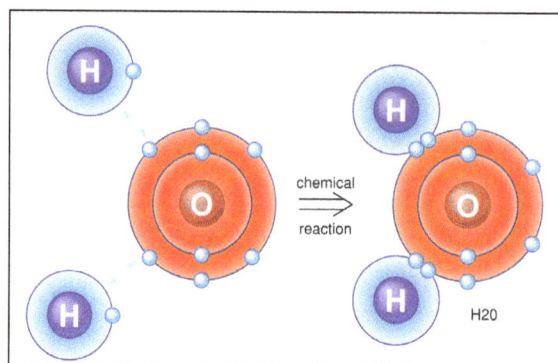

Polar covalent bond

In polar covalent bonds, such as that between hydrogen and oxygen atoms, the electrons are not transferred from one atom to the other as they are in an ionic bond. Instead, some outer electrons merely spend more time in the vicinity of the other atom. The effect of this orbital distortion is to induce regional net charges that hold the atoms together, such as in water molecules.

The electron-sharing, or covalent, bond is the strongest of all chemical bond types. Minerals bonded in this manner display general insolubility, great stability, and a high melting point. Crystals of covalently bonded minerals tend to exhibit lower symmetry than their ionic counterparts because the covalent bond is highly directional, localized in the vicinity of the shared electrons.

The Cl_2 molecules formed by linking two neighbouring chlorine atoms are stable and do not combine with other molecules. Atoms of some elements, however, have more than one electron in the outer orbital and thus may bond to several neighbouring atoms to form groups, which in turn may join together in larger combinations. Carbon, in the polymorphic form of diamond, is a good example of this type of covalent bonding. There are four valence electrons in a carbon atom, so that each atom bonds with four others in a stable tetrahedral configuration. A continuous network is formed by the linkage of every carbon atom in this manner. The rigid diamond structure results from the strong localization of the bond energy in the vicinity of the shared electrons; this makes diamond the hardest of all natural substances. Diamond does not conduct electricity, because all the valence electrons of its constituent atoms are shared to form bonds and therefore are not mobile.

Metallic Bonds

Bonding in metals is distinct from that in their salts, as reflected in the significant differences between the properties of the two groups. In contrast to salts, metals display high plasticity, tenacity, ductility, and conductivity. Many are characterized by lower hardness and have higher melting and boiling points than, for example, covalently bonded materials. All these properties result from a metallic bonding mechanism that can be envisioned as a collection of positively charged ions immersed in a cloud of valence electrons. The attraction between the cations and the electrons holds a crystal together. The electrons are not bound to any particular cation and are thus free to move throughout the structure. In fact, in the metals sodium, cesium, rubidium, and potassium, the radiant energy of light can cause electrons to be removed from their surfaces entirely. (This result is known as the photoelectric effect.) Electron mobility is responsible for the ability of metals to conduct heat and electricity. The native metals are the only minerals to exhibit pure metallic bonding.

Van Der Waals Bonds

Neutral molecules may be held together by a weak electric force known as the van der Waals bond. It results from the distortion of a molecule so that a small positive charge develops on one end and a corresponding negative charge develops on the other. A similar effect is induced in neighbouring molecules, and this dipole effect propagates throughout the entire structure. An attractive force is then formed between oppositely charged ends of the dipoles. Van der Waals bonding is common in gases and organic liquids and solids, but it is rare in minerals. Its presence in a mineral defines a weak area with good cleavage and low hardness. In graphite, carbon atoms lie in covalently bonded sheets with van der Waals forces acting between the layers.

Hydrogen Bonds

There is an interaction called hydrogen bonding. This takes place when a hydrogen atom, bonded to an electronegative atom such as oxygen, fluorine, or nitrogen, is also attracted to the negative end of a neighbouring molecule. A strong dipole-dipole interaction is produced, forming a bond between the two molecules. Hydrogen bonding is common in hydroxides and in many of the layer silicates—e.g., micas and clay minerals.

Physical Properties

The physical properties of minerals are the direct result of the structural and chemical characteristics of the minerals. Some properties can be determined by inspection of a hand specimen or by relatively simple tests on such a specimen. Others, such as those determined by optical and X-ray diffraction techniques require special and often sophisticated equipment and may involve elaborate sample preparation. In the discussion that follows, emphasis is placed on those properties that can be most easily evaluated with only simple tests.

Crystal Habit and Crystal Aggregation

The external shape (habit) of well-developed crystals can be visually studied and classified according to the various crystal systems that span the 32 crystal classes. The majority of crystal occurrences, however, are not part of well-formed single crystals but are found as crystals grown together in aggregates. Examples of some descriptive terms for such aggregations are given here: granular, an intergrowth of mineral grains of approximately the same size; lamellar, flat, platelike individuals arranged in layers; bladed, elongated crystals flattened like a knife blade; fibrous, an aggregate of slender fibres, parallel or radiating; acicular, slender, needlelike crystals; radiating, individuals forming starlike or circular groups; globular, radiating individuals forming small spherical or hemispherical groups; dendritic, in slender divergent branches, somewhat plantlike; mammillary, large smoothly rounded, masses resembling mammae, formed by radiating crystals; botryoidal, globular forms resembling a bunch of grapes; colloform, spherical forms composed of radiating individuals without regard to size (this includes botryoidal, reniform, and mammillary forms); stalactitic, pendant cylinders or cones resembling icicles; concentric, roughly spherical layers arranged about a common centre, as in agate and in geodes; geode, a partially filled rock cavity lined by mineral material (geodes may be banded as in agate owing to successive depositions of material, and the inner surface is often covered with projecting crystals); and oolitic, an assemblage consisting of small spheres.

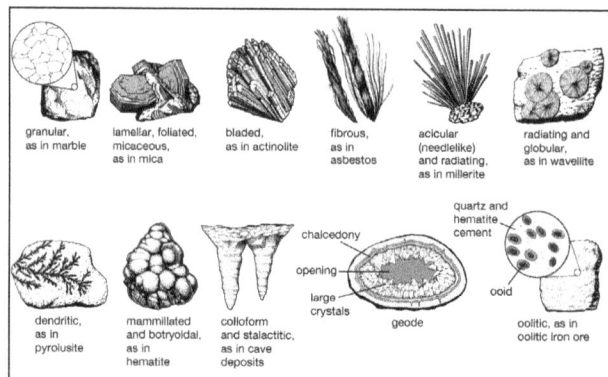

Common crystals: Common crystal aggregations and habits

Cleavage and Fracture

Both these properties represent the reaction of a mineral to an external force. Cleavage is breakage along planar surfaces, which are parallel to possible external faces on the crystal. It results from the tendency of some minerals to split in certain directions that are structurally weaker than others. Some crystals exhibit well-developed cleavage, as seen by the planar cleavage in mica; perfect cleavage of this sort is characterized by smooth, shiny surfaces. In other minerals, such as quartz, cleavage is absent. Quality and direction are the general characteristics used to describe cleavage. Quality is expressed as perfect, good, fair, and so forth; cleavage directions of a crystal are consistent with its overall symmetry.

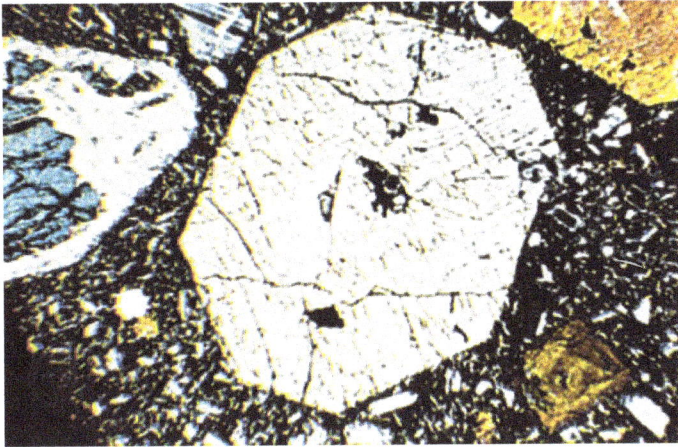

Augite: Photomicrograph of various pyroxene minerals in thin sections (illuminated with polarized light). An augite phenocryst (large individual gray crystal) appears in basalt lava, showing characteristic basal octagonal form and square-segmentation cleavage.

Some crystals do not usually break in any particular direction, reflecting roughly equal bond strengths throughout the crystal structure. Breakage in such minerals is known as fracture. The term conchoidal is used to describe fracture with smooth, curved surfaces that resemble the interior of a seashell; it is commonly observed in quartz and glass. Splintery fracture is breakage into elongated fragments like splinters of wood, while hackly fracture is breakage along jagged surfaces.

Lustre

The term lustre refers to the general appearance of a mineral surface in reflected light. The main types of lustre, metallic and non-metallic, are distinguished easily by the human eye after some practice, but the difference between them cannot be quantified and is rather difficult to describe. Metallic refers to the lustre of an untarnished metallic surface such as gold, silver, copper, or steel. These materials are opaque to light; none passes through even at thin edges. Pyrite (FeS_2), chalcopyrite ($CuFeS_2$), and galena (PbS) are common minerals that have metallic lustre. Nonmetallic lustre is generally exhibited by light-coloured minerals that transmit light, either through thick portions or at least through their edges. The following terms are used to distinguish the lustre of nonmetallic minerals: vitreous, having the lustre of a piece of broken glass (this is commonly seen in quartz and many other nonmetallic minerals); resinous, having the lustre of a piece of resin (this is common in sphalerite [ZnS]); pearly, having the lustre of mother-of-pearl (i.e., an iridescent pearl-like lustre characteristic of mineral surfaces that are parallel to well-developed cleavage planes; the cleavage surface of talc [$Mg_3Si_4O_{10}(OH)_2$] may show pearly lustre); greasy, having the appearance of being covered with a thin layer of oil (such lustre results from the scattering of light by a microscopically rough surface; some nepheline [$(Na, K)AlSiO_4$] and milky quartz may exhibit this); silky, descriptive of the lustre of a skein of silk or a piece of satin and characteristic of some minerals in fibrous aggregates (examples are fibrous gypsum [$CaSO_4 \cdot 2H_2O$], known as satin spar, and chrysotile asbestos [$Mg_3Si_2O_5(OH)_4$]); and adamantine, having the brilliant lustre of diamond, exhibited by minerals with a high refractive index comparable to diamond and which as such refract light as strongly as the latter (examples are cerussite [$PbCO_3$] and anglesite [$PbSO_4$]).

Galena: Galena is the most common mineral that contains lead.

Colour

Minerals occur in a great variety of colours. Because colour varies not only from one mineral to another but also within the same mineral (or mineral group), the observer must learn in which minerals it is a constant property and can thus be relied on as a distinguishing criterion. Most minerals that have a metallic lustre vary little in colour, but nonmetallic minerals can demonstrate wide variance. Although the colour of a freshly broken surface of a metallic mineral is often highly diagnostic, this same mineral may become tarnished with time. Such as tarnish may dull minerals such as galena (PbS), which has a bright bluish lead-gray colour on a fresh surface but may become dull upon long exposure to air. Bornite (Cu_5FeS_4), which on a freshly broken surface has a brownish bronze colour, may be so highly tarnished on an older surface that it shows variegated purples and blues; hence, it is called peacock ore. In other words, in the identification of minerals with a metallic lustre, it is important for the observer to have a freshly broken surface for accurate determination of colour.

Bornite

Aquamarine: Single crystal of aquamarine in matrix

A few minerals with non-metallic lustre display a constant colour that can be used as a truly diagnostic property. Examples are malachite, which is green; azurite, which is blue; rhodonite, which is pink; turquoise, which gives its name to the colour turquoise, a greenish blue to blue-green; and sulfur, which is yellow. Many non-metallic minerals have a relatively narrow range of colours, although some have an unusually wide range. Members of the plagioclase feldspar series range from almost pure white in albite through light gray to darker gray toward the anorthite end-member. Most common garnets show various shades of red to red-brown to brown. Members of the monoclinic pyroxene group range from almost white in pure diopside to light green in diopside containing a small amount of iron as a substitute for magnesium in the structure through dark green in hedenbergite to almost black in many augites. Members of the orthopyroxene series (enstatite to orthoferrosilite) range from light beige to darker brown. On the other hand, tourmaline may show many colours (red, blue, green, brown, and black) as well as distinct colour zonation, from colourless through pink to green, within a single crystal. Similarly, numerous gem minerals such as corundum, beryl, and quartz occur in many colours; the gemstones cut from them are given varietal names. In short, in non-metallic minerals of various kinds, colour is a helpful, though not a truly diagnostic (and therefore unique), property.

Hardness

Hardness (H) is the resistance of a mineral to scratching. It is a property by which minerals may be described relative to a standard scale of 10 minerals known as the Mohs scale of hardness. The degree of hardness is determined by observing the comparative ease or difficulty with which one mineral is scratched by another or by a steel tool. For measuring the hardness of a mineral, several common objects that can be used for scratching are helpful, such as a fingernail, a copper coin, a steel pocketknife, glass plate or window glass, the steel of a needle, and a streak plate (an unglazed black or white porcelain surface).

Mohs hardness scale and observations on hardness of some additional materials			
Mineral	Mohs hardness	Other materials	Observations on the minerals
Talc	1		Very easily scratched by the fingernail; has a greasy feel
Gypsum	2	~2.2 fingernail	Can be scratched by the fingernail
Calcite	3	~3.2 copper penny	Very easily scratched with a knife and just scratched with a copper coin
Fluorite	4		Very easily scratched with a knife but not as easily as calcite
Apatite	5	~5.1 pocketknife	Scratched with a knife with difficulty
		~5.5 glass plate	
Orthoclase	6	~6.5 steel needle	Cannot be scratched with a knife, but scratches glass with difficulty
Quartz	7	~7.0 streak plate	Scratches glass easily
Topaz	8		Scratches glass very easily
Corundum	9		Cuts glass
Diamond	10		Used as a glass cutter

Because there is a general link between hardness and chemical composition, these generalizations can be made:

1. Most hydrous minerals are relatively soft (H < 5).

2. Halides, carbonates, sulfates, and phosphates also are relatively soft (H < 5.5).

3. Most sulfides are relatively soft (H < 5), with marcasite and pyrite being examples of exceptions (H < 6 to 6.5).

4. Most anhydrous oxides and silicates are hard (H > 5.5).

Because hardness is a highly diagnostic property in mineral identification, most determinative tables use relative hardness as a sorting parameter.

Tenacity

Several mineral properties that depend on the cohesive force between atoms (and ions) in mineral structures are grouped under tenacity. A mineral's tenacity can be described by the following terms: malleable, capable of being flattened under the blows of a hammer into thin sheets without breaking or crumbling into fragments (most of the native elements show various degrees of malleability, but particularly gold, silver, and copper); sectile, capable of being severed by the smooth cut of a knife (copper, silver, and gold are sectile); ductile, capable of being drawn into the form of a wire (gold, silver, and copper exhibit this property); flexible, bending easily and staying bent after the pressure is removed (talc is flexible); brittle, showing little or no resistance to breakage, and as such separating into fragments under the blow of a hammer or when cut by a knife (most silicate minerals are brittle); and elastic, capable of being bent or pulled out of shape but returning to the original form when relieved (mica is elastic).

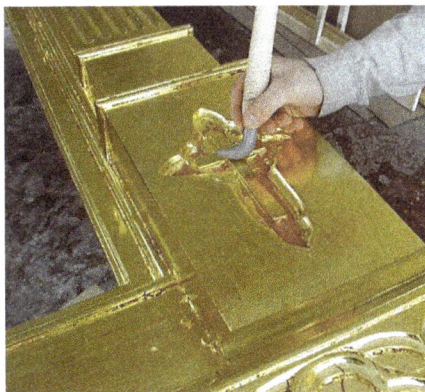

Gold leaf: Burnishing gold leaf.

Specific Gravity

Specific gravity (G) is defined as the ratio between the weight of a substance and the weight of an equal volume of water at 4 °C (39 °F). Thus a mineral with a specific gravity of 2 weighs twice as much as the same volume of water. Since it is a ratio, specific gravity has no units.

The specific gravity of a mineral depends on the atomic weights of all its constituent elements and the manner in which the atoms (and ions) are packed together. In mineral series whose species have essentially identical structures, those composed of elements with higher atomic weight have higher specific gravities. If two minerals (as in the two polymorphs of carbon, namely graphite and diamond) have the same chemical composition, the difference in specific gravity reflects variation in internal packing of the atoms or ions (diamond, with a G of 3.51, has a more densely packed structure than graphite, with a G of 2.23).

Measurement of the specific gravity of a mineral specimen requires the use of a special apparatus. An estimate of the value, however, can be obtained by simply testing how heavy a specimen feels. Most people, from everyday experience, have developed a sense of relative weights for even such objects as non-metallic and metallic minerals. For example, borax (G = 1.7) seems light for a non-metallic mineral, whereas anglesite (G = 6.4) feels heavy. Average specific gravity reflects what a non-metallic or metallic mineral of a given size should weigh. The average specific gravity for non-metallic minerals falls between 2.65 and 2.75, which is seen in the range of values for quartz (G = 2.65), feldspar (G = 2.60 to 2.75), and calcite (G = 2.72). For metallic minerals, graphite (G = 2.23) feels light, while silver (G = 10.5) seems heavy. The average specific gravity for metallic minerals is approximately 5.0, the value for pyrite. With practice using specimens of known specific gravity, a person can develop the ability to distinguish between minerals that have comparatively small differences in specific gravity by merely lifting them.

Although an approximate assessment of specific gravity can be obtained by the hefting of a hand specimen of a specific monomineral, an accurate measurement can only be achieved by using a specific gravity balance. An example of such an instrument is the Jolly balance, which provides numerical values for a small mineral specimen (or fragment) in air as well as in water. Such accurate measurements are highly diagnostic and can greatly aid in the identification of an unknown mineral sample.

Magnetism

Only two minerals exhibit readily observed magnetism: magnetite (Fe_3O_4), which is strongly attracted to a hand magnet, and pyrrhotite ($Fe_{1-x}S$), which typically shows a weaker magnetic reaction. Ferromagnetic is a term that refers to materials that exhibit strong magnetic attraction when subjected to a magnetic field. Materials that show only a weak magnetic response in a strong applied magnetic field are known as paramagnetic. Those materials that are repelled by an applied magnetic force are known as diamagnetic. Because minerals display a wide range of slightly different magnetic properties, they can be separated from each other by an electromagnet. Such magnetic separation is a common procedure both in the laboratory and on a commercial scale.

Fluorescence

Some minerals, when exposed to ultraviolet light, will emit visible light during irradiation; this is known as fluorescence. Some minerals fluoresce only in shortwave ultraviolet light, others only in long wave ultraviolet light, and still others in either situation. Both the colour and intensity of the emitted light vary significantly with the wavelengths of ultraviolet light. Due to the unpredictable nature of fluorescence, some specimens of a mineral manifest it, while other seemingly similar specimens, even those from the same geographic area, do not. Some minerals that may exhibit fluorescence are fluorite, scheelite, calcite, scapolite, willemite and autunite. Specimens of willemite and calcite from the Franklin district of New Jersey in the United States may show brilliant fluorescent colours.

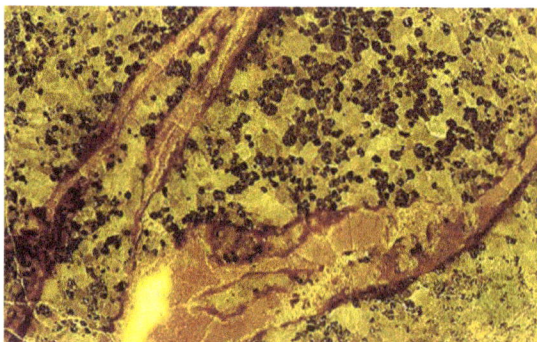

Franklinite and willemite: In ordinary light, the willemite grains and veins of this specimen are brownish gold and the franklinite grains are dark brown.

Solubility in Hydrochloric Acid

The positive identification of carbonate minerals is aided greatly by the fact that the carbon-oxygen bond of the CO_3 group in carbonates becomes unstable and breaks down in the presence of hydrogen ions (H^+) available in acids. This is expressed by the reaction $2H^+ + CO_3^- \longrightarrow H_2O + CO_2$, which is the basis for the so-called fizz test with dilute hydrochloric acid (HCl). Calcite, aragonite, witherite, and strontianite, as well as copper carbonates, show bubbling, or effervescence, when a drop of dilute hydrochloric acid is placed on the mineral. This "fizz" is due to the release of carbon dioxide (CO_2). Other carbonates such as dolomite, rhodochrosite, magnesite, and siderite will show slow effervescence when acid is applied to powdered minerals or moderate effervescence only in hot hydrochloric acid.

Radioactivity

Minerals containing uranium (U) and thorium (Th) continually undergo decay reactions in which radioactive isotopes of uranium and thorium form various daughter elements and also release energy in the form of alpha and beta particles and gamma radiation. The radiation produced can be measured in the laboratory or in the field using a Geiger counter or a scintillation counter. A radiation counter therefore is helpful in identifying uranium- and thorium-containing minerals, such as uraninite, pitchblende, thorianite, and autunite. Several rock-forming minerals contain enough radioactive elements to permit the determination of the time elapsed since the radioactive material was incorporated into the mineral.

Properties of Minerals

Physical Properties

Color, luster, streak, hardness, cleavage, fracture and crystal form are the most useful physical properties for identifying most minerals. Other properties-such as reaction with acid, magnetism, specific gravity, tenacity, taste, odor, feel, and presence of striations-are helpful in identifying certain minerals.

Luster

Examples of mineral luster:
Waxy (chalcedony); Pearly (talc); and Earthy (goethite).

Luster describes the appearance of a mineral when light is reflected from its surface. Is it shiny or dull: does it look like a metal or like glass? Generally the first thing you notice when identifying an unknown sample is the mineral's luster. Important examples of mineral luster are shown in figure.

Minerals with a metallic luster look like a metal, such as steel or copper. They are both shiny and opaque, even when looking at a thin edge. Many metallic minerals become dull or earthy when they are exposed to the elements for a long time (like silver, they tarnish). To determine whether or not a mineral has a metallic luster, therefore, you must look at a recently broken part of the mineral.

Minerals with an earthy luster look like earth, or dirt. Like metallic minerals these are completely opaque, but dull. Again, think of rust on iron or tarnish that forms on precious metals.

Vitreous luster- is like that of glass, shiny and translucent to transparent. Remember that glass can be almost any color, including black, so don't be fooled by the color. Also, a dark piece of glass may appear to be opaque if it is thick enough. If you hold a thin edge up to the light you should be able to see light bleeding through.

Minerals with a waxy luster look like paraffin, typically translucent but dull. While minerals with pearly luster have an appearance similar to a pearl or the inside of an abalone shell – translucent and shiny but with a bit of light refraction, producing a rainbow effect on the surface (similar to an oil slick).

Color

Color is one of the most obvious properties of a mineral but it is often of limited diagnostic value, especially in minerals that are not opaque. While many metallic and earthy minerals have distinctive colors, transucent or transparent minerals can vary widely in color. Quartz, for example, can vary from colorless to white to yellow to gray to pink to purple to black. On the other hand the colors of some minerals, such as biotite (black) and olivine (olive green) can be distinctive. Never use color as a final diagnostic property -- check other properties before making an identification.

Quartz varies widely in color, due to minor (parts per billion) impurities and even defects in its crystalline structure. Pure quartz has no color, as shown at left. Colored varieties of quartz, above from left to right, include: amethyst (purple); citrine (yellow); rose (pink); rock crystal (clear).

Streak

Streak refers to the color of the mineral in its powdered form, which may or may not be the same color as the mineral. Streak is helpful for identifying minerals with metallic or earthy luster, because minerals with nonmetallic luster generally have a colorless or white streak that is not diagnostic, Streak is obtained by scratching the mineral on an unpolished piece of white porcelain called a streak plate. Because the streak plate is harder than most minerals, rubbing the mineral across the plate produces a powder of that mineral. When the excess powder is blown away, what remains is the color of the streak. Because the streak of a mineral is usually the same, no matter what the color of the mineral, streak is commonly more reliable than color for identification.

The streak of this dark gray mineral (hematite), obtained by rubbing it on the white streak plate is reddish brown.

Hardness

Hardness is the resistance of a mineral to scratching or abrasion by other materials. Hardness is determined by scratching the surface of the sample with another mineral or material of known hardness. The standard hardness scale, called Mohs Hardness Scale, consists of ten minerals ranked in ascending order of hardness with diamond, the hardest known substance, assigned the number 10. The hardness kits we use in class contain only minerals 2-7, as these are the most useful for testing most of the minerals we will encounter in this class. Since most of us don't wander the outdoors with a pocketful of standard minerals table one also lists the relative hardness of other common items.

Table: Relative Hardness

Mohs Scale of Hardness		Common Objects (hardness)	
1.	Talc		
2.	Gypsum	Fingernail	(2.5)

3.	Calcite	Copper penny	(3-3.5)
4.	Fluorite		
5.	Apatite	Glass	(5.5)
6.	Feldspar	Steel nail or knife blade	(6-6.5)
7.	Quartz		
8.	Topaz		
9.	Corundum		
10.	Diamond		

To determine hardness, run a sharp edge or a point of a mineral with known hardness across a smooth face of the mineral to be tested. Do not scratch back and forth like an eraser, but press hard and slowly scratch a line, like you are trying to etch a groove in glass. Sometimes powder of the softer mineral is left on the harder mineral and gives the appearance of a scratch on the harder one. Brush the tested surface with your finger to see if a groove or scratch remains.

A piece of glass is provided in the hardness kits as a standard for determining hardness. There are several reasons for this:

• It is easy to see a scratch on glass;

• The hardness of glass (5 to 5½) is midway on the mohs scale; and

• Glass is inexpensive and easily replaced.

Testing mineral hardness. The harder mineral (quartz) scratches the softer one (calcite).

The hardness of a mineral when compared to glass is one of the principal bases in identifying a mineral. Put the piece of glass on a stable, flat surface such as a table top. Drag a corner or edge of the mineral across the surface of the glass and check to see if the glass was scratched.

The Mohs scale was developed by Friedrich Mohs in 1822 as a diagnostic tool. One measure of absolute hardness is an indentation test in which a diamond-tipped tool is impressed into a sample with a fixed amount of pressure and the depth of the resulting groove is measured. The figure below shows a graph comparing this numerical hardness (called the Knoop hardness) to Mohs hardness.

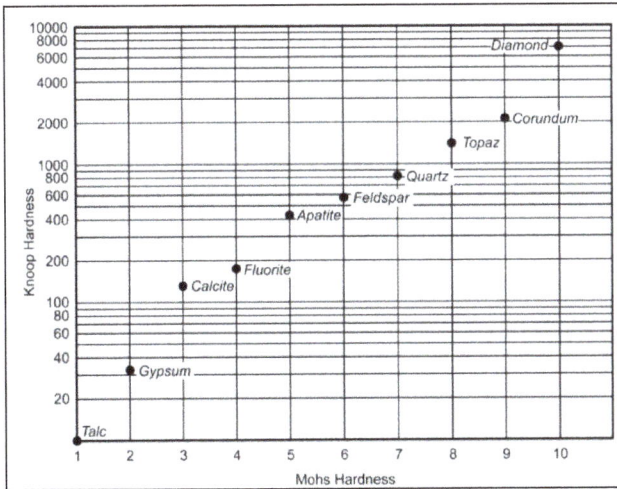

Comparison between Mohs scale hardness and Knoop hardness.
The Knoop hardness is shown with a logarithmic scale.

Cleavage and Fracture

The way in which a mineral breaks is determined by the arrangement of its atoms and the strength of the chemical bonds holding them together. Because these properties are unique to the mineral, careful observation of broken surfaces may aid in mineral identification. A mineral that exhibits cleavage consistently breaks, or cleaves, along parallel flat surfaces called cleavage planes. A mineral fractures if it breaks along random, irregular surfaces. Some minerals break only by fracturing, while others both cleave and fracture.

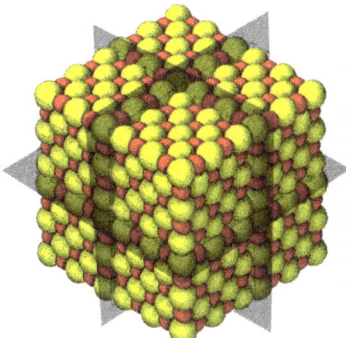

The illustration shows that atoms of sodium (red) and chlorine (yellow) in the mineral halite are parallel to three planes that intersect at 90°. Halite breaks, or cleaves, most easily between the three planes of atoms, so it has three directions of cleavage that intersect at 90°. The photograph illustrates the three directions of cleavage in halite.

The mineral halite (NaCl or sodium chloride) illustrates how atomic arrangement determines the way a mineral breaks. Figure shows the arrangement of sodium and chlorine atoms in halite. Notice that there are planes with atoms and planes without atoms. When halite breaks, it breaks parallel to the planes with atoms but along the planes without atoms. Because there are three directions in which atom density is equal, halite has three directions of cleavage, each at 90° to each other. The number of cleavage directions and the angles between them are important in mineral identification because they reflect the underlying atomic architecture that defines each mineral.

Cleavage planes, as flat surfaces, are easily spotted by turning a sample in your hand until you see a single flash of reflected light from across the mineral surface. Individual cleavage surfaces may extend across the whole mineral specimen or, more commonly, they may be offset from each other by small amounts, as illustrated in figure. Even though they are offset, they work as tiny mirrors that create the single flash seen in figure. Figures also show offset cleavage surfaces. Cleavage quality is described as perfect, good, and poor. Minerals with a perfect or excellent cleavage break easily along flat surfaces and are easy to spot. Minerals with good cleavages do not have such well-defined cleavage planes and reflect less light. Poor- cleavages are the toughest to recognize, but can be spotted by small flashes of light in certain positions. All cleavages illustrated here are perfect or good.

The above figure shows that when incoming light rays (yellow lines) strike parallel cleavage surfaces, the rays are all reflected in the same direction (red lines), even though the surfaces are at different elevations. When viewed from that direction, the mineral shines. Rays striking irregular fracture surfaces reflect in many different directions (black lines). Causing fracture surfaces to appear duller. The parallel cleavage planes in hornblende shine because the light is reflected in the same direction.

Minerals have characteristic numbers of cleavages. This number is determined by counting the number of cleavage surfaces that are not parallel to each other. For example, the mineral in figure has two planes of cleavage, one that is visible and one lying on the table. However, each of these cleavage surfaces is parallel to the other, so this mineral is said to have only one cleavage direction. Minerals with one cleavage are often said to have a basal cleavage.

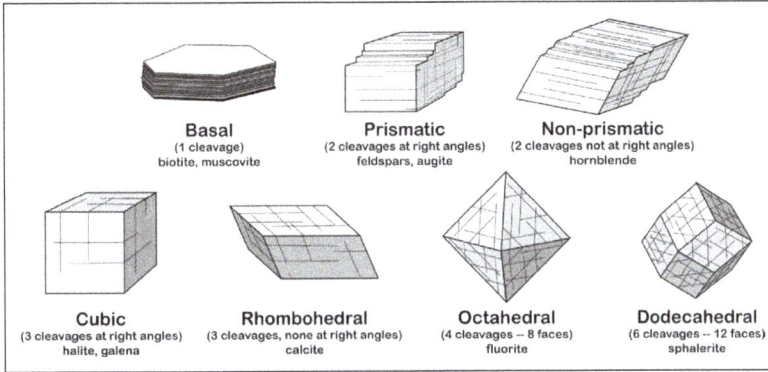

Types of cleavage common in minerals with examples of minerals.

The mineral biotite has basal cleavage, meaning it has one perfect cleavage. The cleavage plane
is apparent in the flat, reflective surface on top of this sample. The flat surface on the bottom, in contact
with the table top, is parallel to the top and therefore represents the same plane of cleavage.

Two cleavage directions are present when planes of breakage occur along two non-parallel planes. These two planes can be perpendicular (at 90°) to one another, in which case the mineral is said to have prismatic cleavage. In some minerals the two planes of cleavage may not be perpendicular – this is known as non-prismatic cleavage. When there are two cleavages, you should note the angle between them. Most commonly, cleavage angles are close to 60°, 90°, or 120°.

Two directions of cleavage: A. Cleavages in potassium feldspar intersect at 90°;
B. Prismatic cleavages in hornblende intersect at 56° and 124°.

Some minerals have three planes of cleavage: If the three cleavages intersect at 90° the mineral is said to have cubic cleavage: If none of the cleavage planes intersect at right angles the shape is a squashed cube known as a rhombohedron. A third variant occurs when a mineral has two cleavage planes that are perpendicular, and a third that is not perpendicular to the other two.

Three directions of cleavage - A. Cleavages in halite intersect at 90° (cubic cleavage);
B. Cleavages do not intersect at 90° in calcite (rhombohedral cleavage);
C. Two of the cleavage planes in gypsum intersect at 90°.

Minerals with four or six cleavage directions are not common. Four cleavage planes can intersect to form an eight sided figure known as an octahedron. Fluorite is the most common mineral with an octahedral cleavage. Six cleavage directions intersect to form a dodecahedron, a twelve-sided form with diamond-shaped faces. A common mineral with dodecahedal cleavage is sphalerite.

Fluorite (left) has four perfect cleavages; this octahedron is made up of eight equilateral triangular faces. The mineral sphalerite has six directions of cleavage (dodecahedral cleavage).

When counting cleavage directions it is essential that you count surfaces on just one mineral crystal. The photographs shown here used single large, broken crystals to illustrate cleavage. In nature you often find that a single hand-sized sample contains a large number of crystals grown together. If you count cleavage surfaces from more than one crystal, a wrong number is likely.

Conchoidal fracture in quartz (left) and Chalcedony (right). This type of fracture once made flint (black chalcedony) and obsidian (volcanic glass, which also displays conchoidal fracture) valuable for making sharp stone tools.

Finally, fracture surfaces can cut a mineral grain in any direction. Fractures are generally rough or irregular, rather than flat, and thus appear duller than cleavage surfaces.

Some minerals fracture in a way that helps to identify them. For example, quartz has no cleavage but, like glass, it breaks along numerous small, smooth, curved surfaces called conchoidal fractures. Though other kinds of fracture exist in nature, such as fibrous, splintery, or irregular, conchoidal fracture is the only type we will concern ourselves with here.

In the field you will often have to break samples into pieces to observe cleavages and fractures on fresh surfaces. While it is instructional to hammer some mineral samples yourself, do not break the lab samples without your instructor's approval. Samples cost money and in most cases have already been broken to show characteristic features.

Crystal Form

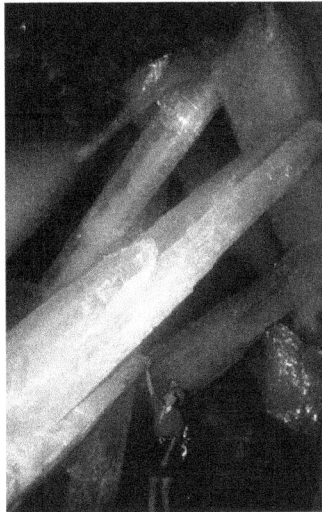

A visitor stands among a cluster of giant prismatic gypsum crystals
discovered in a section of the Naica lead-zinc mine in Chihuahua, Mexico. Some of the
individual crystals here are up to 10 meters in length.

A crystal is a solid, homogeneous, orderly array of atoms and may be nearly any size. The arrangement of atoms within a mineral determines the external shape of its crystals. Some crystals have smooth, planar faces and regular, geometric shapes; these are what most people think of as crystals. These crystals occur only rarely in nature however, because in order to develop those beautifully-shaped faces the mineral must have unlimited space in which to grow.

When a mineral begins to solidify, either due to the cooling of molten material or due to precipitation from a solution, microscopic crystals always form and grow. These tiny crystals will continue to grow until they run out of space. At this point their external shape will simply reflect the shape of the void which they grew. If the growing crystal runs out of material before it runs out of space, you will be left with a nicely shaped crystal within an otherwise empty void such as a geode.

Crystal of quartz Sides form a hexagonal prism that is capped with pyramid-like faces. Note the fine grooves on some crystal faces.

Crystals of pyrite in the form of a cube (left) and pyritohedrons (right). Note the fine grooves on the faces of these crystals.

Some minerals commonly occur as well-developed crystals, and their crystal forms are diagnostic. A detailed nomenclature has evolved to describe crystal forms, and some of the common names may be familiar For example, quartz commonly occur as hexagonal (six-sided) prisms with pyramid-like shapes at the top. Pyrite occurs as cubes or pyritohedrons (forms with twelve pentagonal faces). Garnets occur as dodecahedrons, 12-sided forms that have a roughly round shape.

Garnets usually form dodecahedral (12-sided) crystals with rhombic faces. Note that the color of garnet may vary widely, from the reddish-brown pyropes in the image on the left to the whitegreen grossular garnets on the right.

Cleavage surfaces may be confused with natural crystal faces; in fact, cleavage planes are parallel to possible (but not always developed) crystal faces. They can be distinguished as follows:

1. Crystal faces are normally smooth, whereas cleavage planes, though also smooth, commonly are broken in a step-like fashion.

2. Some crystal faces have fine grooves or ridges on their surfaces whereas cleavage planes do not. Similar looking, very thin, parallel grooves, or striations, are seen on plagioclase cleavage surfaces, but these features persist throughout the mineral and are not surficial.

3. Finally, unless crystal faces happen to coincide with cleavage planes, the mineral will not break parallel to them.

Additional Properties

Special properties help identify some minerals. These properties may not be distinctive enough in most minerals to help with their identification or they may be present only in certain minerals.

Magnetism

Some minerals are attracted to a hand magnet. To test a mineral for magnetism, just put the magnet and mineral together and see if they are attracted. Magnetite is the only common mineral that is always strongly magnetic.

Reaction with Acid

Some minerals especially carbonate minerals, react visibly with acid. (Usually, a dilute hydrochloric acid [HCl] is used.) When a drop of dilute hydrochloric acid is placed on calcite, it readily bubbles or effervesces, releasing carbon dioxide. BE CAREFUL when using the acid - dilute acid can burn your skin (especially if you have a cut) or stain or put a hole in your clothing. Only a small drop of acid is needed to see whether or not the mineral bubbles. When you finish making the test, wipe the acid off the mineral immediately. Should you get acid on yourself, wash it off right away; if you get it on your clothing, rinse it out immediately.

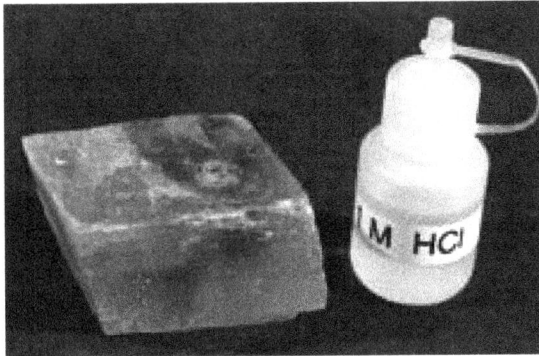

Calcite reacting to a drop of acid.

Striations

Plagioclase feldspar can be positively identified and distinguished from potassium feldspar by the presence of very thin, parallel grooves called striations. The grooves are present on only one of the two sets of cleavages and are best seen with a hand lens. They may not be visible on all parts of a cleavage surface. Before you decide there are no striations, look at all parts of all visible cleavage surfaces, moving the sample around as you look so that light is reflected from these surfaces at different angles.

Until you have seen striations for the first time, you may confuse them with the small, somewhat irregular, differently colored intergrowths or veinlets seen on cleavage faces of some specimens of potassium feldspar. However, these have variable widths, are not strictly parallel, and are not grooves, so they are easily distinguished from striations.

The thin veinlets seen in some potassium feldspars (left) should not be confused with striations in plagioclase (right). Striations are visible on the upper surface of this sample of plagioclase.

Specific Gravity

The specific gravity of a mineral is the weight of that mineral divided by the weight of an equal volume of water. The specific gravity of water equals 1.0, by definition. Most silicate, or rock-forming, minerals have specific gravities of 2.6 to 3.4; the ore minerals are usually heavier, with specific gravities of 5 to 8. If you compare similarsized samples of two different minerals, the one with the higher specific gravity will feel the heaviest; it has a greater heft. For most minerals, specific gravity is not a particularly noteworthy feature, but for some, high specific gravity is distinctive (examples are barite and galena).

Taste, Odor and Feel

Some minerals have a distinctive taste (halite is salt, and tastes like it). Some a distinctive odor (the powder of some sulfide minerals, such as sphalerite, a zinc sulfide, smells like rotten eggs), and some a distinctive feel (talc feels slippery).

Classification of Minerals

Since the middle of the 19th century, minerals have been classified on the basis of their chemical composition. Under this scheme, they are divided into classes according to their dominant anion or anionic group (e.g., halides, oxides, and sulfides). Several reasons justify use of this criterion as the distinguishing factor at the highest level of mineral classification. First, the similarities in properties of minerals with identical anionic groups are generally more pronounced than those with the same dominant cation. For example, carbonates have stronger resemblance to one another than do copper minerals. Secondly, minerals that have identical dominant anions are likely to be found in the same or similar geologic environments. Therefore, sulfides tend to occur together in vein or replacement deposits, while silicate-bearing rocks make up much of Earth's crust. Third, current chemical practice employs a nomenclature and classification scheme for inorganic compounds based on similar principles.

Investigators have found, however, that chemical composition alone is insufficient for classifying minerals. Determination of internal structures, accomplished through the use of X rays, allows a more complete appreciation of the nature of minerals. Chemical composition and internal structure together constitute the essence of a mineral and determine its physical properties; thus, classification should rely on both. Crystallo chemical principles—i.e., those relating to both chemical composition and crystal structure—were first applied by the British physicist W. Lawrence Bragg and the Norwegian mineralogist Victor Moritz Goldschmidt in the study of silicate minerals. The silicate group was subdivided in part on the basis of composition but mainly according to internal structure. Based on the topology of the SiO_4 tetrahedrons, the subclasses include framework, chain, and sheet silicates, among others. Such mineral classifications are logical and well-defined.

The broadest divisions of the classification are (1) native elements, (2) sulfides, (3) sulfosalts, (4) oxides and hydroxides, (5) halides, (6) carbonates, (7) nitrates, (8) borates, (9) sulfates, (10) phosphates, and (11) silicates.

Native Elements

Apart from the free gases in Earth's atmosphere, some 20 elements occur in nature in a pure (i.e., uncombined) or nearly pure form. Known as the native elements, they are partitioned into three families: metals, semimetals, and non-metals. The most common native metals, which are characterized by simple crystal structures, make up three groups: the gold group, consisting of gold, silver, copper, and lead; the platinum group, composed of platinum, palladium, iridium, and osmium; and the iron group, containing iron and nickel-iron. Mercury, tantalum, tin, and zinc are other metals that have been found in the native state. The native semimetals are divided into two isostructural groups: (1) antimony, arsenic, and bismuth, with the latter two being more common in nature, and (2) the rather uncommon selenium and tellurium. Carbon, in the form of diamond and graphite, and sulfur are the most important native nonmetals.

Native elements	
Metals	
Gold group	
Gold	Au
Silver	Ag
Copper	Cu
Platinum group	
Platinum	Pt
Iron group	

Iron	Fe
(kamacite	Fe, Ni)
(taenite	Fe, Ni)
Semimetals	
Arsenic group	
Arsenic	As
Bismuth	Bi
Non-metals	
Sulfur	S
Diamond	C
Graphite	C

Metals

Gold, silver, and copper are members of the same group (column) in the periodic table of elements and therefore have similar chemical properties. In the uncombined state, their atoms are joined by the fairly weak metallic bond. These minerals share a common structure type, and their atoms are positioned in a simple cubic closest-packed arrangement. Gold and silver both have an atomic radius of 1.44 angstroms (Å), or 1.44×10^{-7} millimetre, which enables complete solid solution to take place between them. The radius of copper is significantly smaller (1.28 Å), and as such copper substitutes only to a limited extent in gold and silver. Likewise, native copper contains only trace amounts of gold and silver in its structure.

Ore: A sample of gold ore

Because of their similar crystal structure, the members of the gold group display similar physical properties. All are rather soft, ductile (capable of being drawn into wire), malleable (capable of being shaped by a hammer or rollers), and sectile (capable of being cut smoothly by a knife or other instrument); gold, silver, and copper serve as excellent conductors of electricity and heat and exhibit metallic lustre and hackly fracture (a type of fracture characterized by sharp jagged surfaces). These properties are attributable to their metallic bonding. The gold-group minerals crystallize in the isometric system and have high densities as a consequence of cubic closest packing.

In addition to the elements listed above, the platinum group also includes rare mineral alloys such as iridosmine. The members of this group are harder than the metals of the gold group and also have higher melting points.

The iron-group metals are isometric and have a simple cubic packed structure. Its members include pure iron, which is rarely found on the surface of Earth, and two species of nickel-iron (kamacite and taenite), which have been identified as common constituents of meteorites. Native iron has been found in basalts of Disko Island, Greenland and nickel-iron in Josephine and Jackson counties, Oregon. The atomic radii of iron and nickel are both approximately 1.24 Å, and so nickel is a frequent substitute for iron. Earth's core is thought to be composed largely of such iron-nickel alloys.

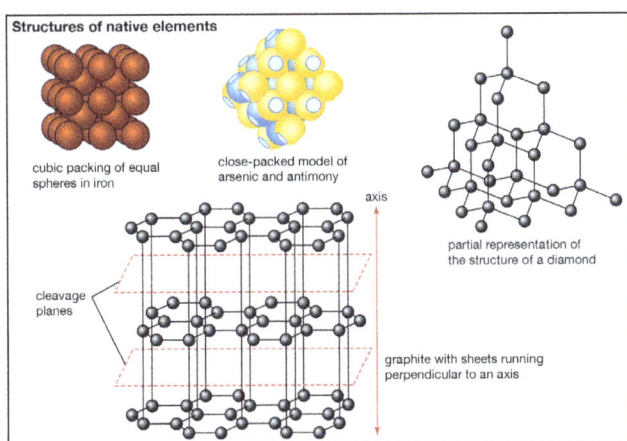

The figure shows the Structures of some native elements. (A) Close-packed model of simple cubic packing of equal spheres, as shown by iron. Each sphere is surrounded by eight closest neighbours. (B) Close-packed model of the structure of arsenic and antimony. Flat areas represent overlap between adjoining atoms. (C) Partial representation of the structure of diamond. (D) The structure of graphite with sheets perpendicular to the c axis.

Semimetals

The semimetals antimony, arsenic, and bismuth have a structure type distinct from the simple-packed spheres of the metals. In these semimetals, each atom is positioned closer to three of its neighbouring atoms than to the rest. The structure of antimony and arsenic is composed of spheres that intersect along flat circular areas.

The covalent character of the bonds joining the four closest atoms is linked to the electronegative nature of the semimetals, reflected by their position in the periodic table. Members of this group are fairly brittle, and they do not conduct heat and electricity nearly as well as the native metals. The bond type suggested by these properties is intermediate between metallic and covalent; it is consequently stronger and more directional than pure metallic bonding, resulting in crystals of lower symmetry.

Arsenic (gray) with realgar (red) and orpiment (yellow)

Nonmetals

The native nonmetals diamond, fullerene, graphite, and sulfur are structurally distinct from the metals and semimetals. The structure of sulfur (atomic radius = 1.04 Å), usually orthorhombic in form, may contain limited solid solution by selenium (atomic radius = 1.16 Å).

The Hope Diamond

The polymorphs of carbon—graphite, fullerene, and diamond—display dissimilar structures, resulting in their differences in hardness and specific gravity. In diamond, each carbon atom is bonded covalently in a tetrahedral arrangement, producing a strongly bonded and exceedingly close-knit but not closest-packed structure. The carbon atoms of graphite, however, are arranged in six-membered rings in which each atom is surrounded by three close-by neighbours located at the vertices of an equilateral triangle. The rings are linked to form sheets, called graphene, that are separated by a distance exceeding one atomic diameter. Van der Waals forces act perpendicular to the sheets, offering a weak bond, which, in combination with the wide spacing, leads to perfect basal cleavage and easy gliding along the sheets. Fullerenes are found in meta-anthracite, in fulgurites, and in clays from the Cretaceous-Tertiary boundary in New Zealand, Spain, and Turkmenistan as well as in organic-rich layers near the Sudbury nickel mine of Canada.

Sulfides

This important class includes most of the ore minerals. The similar but rarer sulfarsenides are grouped here as well. Sulfide minerals consist of one or more metals combined with sulfur; sulfarsenides contain arsenic replacing some of the sulfur.

Sulfides are generally opaque and exhibit distinguishing colours and streaks. (Streak is the colour of a mineral's powder.) The nonopaque varieties (e.g., cinnabar, realgar, and orpiment) possess high refractive indices, transmitting light only on the thin edges of a specimen.

Few broad generalizations can be made about the structures of sulfides, although these minerals can be classified into smaller groups according to similarities in structure. Ionic and covalent bonding are found in many sulfides, while metallic bonding is apparent in others as evidenced by their metal properties. The simplest and most symmetric sulfide structure is based on the architecture of the sodium chloride structure. A common sulfide mineral that crystallizes in this manner is the ore mineral of lead, galena. Its highly symmetric form consists of cubes modified by octahedral faces at their corners. The structure of the common sulfide pyrite (FeS_2) also is modeled after the sodium chloride type; a disulfide grouping is located in a position of coordination with six surrounding ferrous iron atoms. The high symmetry of this structure is reflected in the external morphology of pyrite. In another sulfide structure, sphalerite (ZnS), each zinc atom is surrounded by four sulfur atoms in a tetrahedral coordinating arrangement. In a derivative of this structure type, the chalcopyrite ($CuFeS_2$) structure, copper and iron ions can be thought of as having been regularly substituted in the zinc positions of the original sphalerite atomic arrangement.

Arsenopyrite (FeAsS) is a common sulfarsenide that occurs in many ore deposits. It is the chief source of the element arsenic.

Sulfosalts

There are approximately 100 species constituting the rather large and very diverse sulfosalt class of minerals. The sulfosalts differ notably from the sulfides and sulfarsenides with regard to the role of semimetals, such as arsenic (As) and antimony (Sb), in their structures. In the sulfarsenides, the semimetals substitute for some of the sulfur in the structure, while in the sulfosalts they are found instead in the metal site. For example, in the sulfarsenide arsenopyrite (FeAsS), the arsenic replaces sulfur in a marcasite-(FeS_2-) type structure. In contrast, the sulfosalt enargite (Cu_3AsS_4) contains arsenic in the metal position, coordinated to four sulfur atoms. A sulfosalt such as Cu_3AsS_4 may also be thought of as a double sulfide, $3Cu_2S \cdot As_2S_5$.

Oxides and Hydroxides

These classes consist of oxygen-bearing minerals; the oxides combine oxygen with one or more metals, while the hydroxides are characterized by hydroxyl $(OH)^-$ groups.

The oxides are further divided into two main types: simple and multiple. Simple oxides contain a single metal combined with oxygen in one of several possible metal:oxygen ratios (X:O): XO, X_2O, X_2O_3, etc. Ice, H_2O, is a simple oxide of the X_2O type that incorporates hydrogen as the cation. Although SiO_2 (quartz and its polymorphs) is the most commonly occurring oxide, Two nonequivalent metal sites (X and Y) characterize multiple oxides, which have the form XY_2O_4.

Unlike the minerals of the sulfide class, which exhibit ionic, covalent, and metallic bonding, oxide minerals generally display strong ionic bonding. They are relatively hard, dense, and refractory.

Oxides generally occur in small amounts in igneous and metamorphic rocks and also as preexisting grains in sedimentary rocks. Several oxides have great economic value, including the principal ores of iron (hematite and magnetite), chromium (chromite), manganese (pyrolusite, as well as the hydroxides, manganite and romanechite), tin (cassiterite), and uranium (uraninite).

Members of the hematite group are of the X_2O_3 type and have structures based on hexagonal closest packing of the oxygen atoms with octahedrally coordinated (surrounded by and bonded to six atoms) cations between them. Corundum and hematite share a common hexagonal architecture. In the ilmenite structure, iron and titanium occupy alternate Fe-O and Ti-O layers.

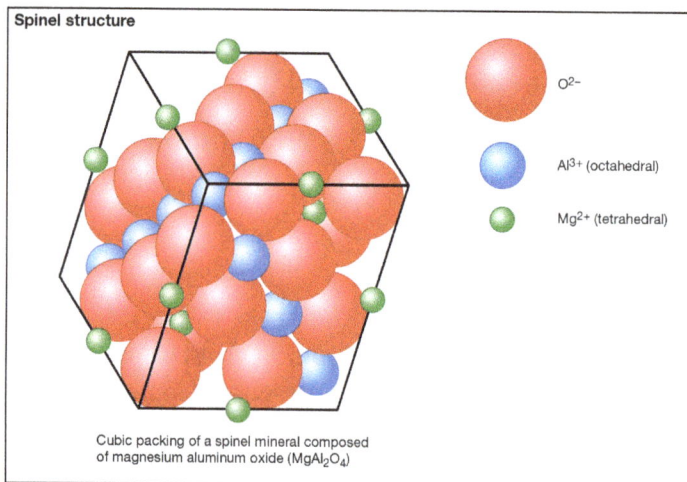

Spinel structure

O^{2-}

Al^{3+} (octahedral)

Mg^{2+} (tetrahedral)

Cubic packing of a spinel mineral composed of magnesium aluminum oxide ($MgAl_2O_4$)

An oxygen layer in the spinel ($MgAl_2O_4$) structure. The large circles represent oxygen in approximate cubic closest packing; the cation layers on each side of the oxygen layer are also shown.

The XO_2-type oxides are divided into two groups. The first structure type, exemplified by rutile, contains cations in octahedral coordination with oxygen. The second resembles fluorite (CaF_2); Each oxygen is bonded to four cations located at the corners of a fairly regular tetrahedron, and each cation lies within a cube at whose corners are eight oxygen atoms. This latter structure is exhibited by uranium, thorium, and cerium oxides, whose considerable importance arises from their roles in nuclear chemistry.

The spinel-group minerals have type XY_2O_4 and contain oxygen atoms in approximate cubic closest packing. The cations located within the oxygen framework are octahedrally and tetrahedrally coordinated with oxygen.

The $(OH)^-$ group of the hydroxides generally results in structures with lower bond strengths than in the oxide minerals. The hydroxide minerals tend to be less dense than the oxides and also are not as hard. All hydroxides form at low temperatures and are found predominantly as weathering products, as, for example, from alteration in hydrothermal veins. Some common hydroxides are brucite $[Mg(OH)_2]$, manganite $[MnO \cdot OH]$, diaspore $[\alpha\text{-AlO} \cdot OH]$, and goethite $[\alpha\text{-FeO} \cdot OH]$. The ore of aluminum, bauxite, consists of a mixture of diaspore, boehmite (γ-AlO $\cdot$ OH—a polymorph of diaspore), and gibbsite $[Al(OH)_3]$, plus iron oxides. Goethite is a common alteration product of iron-rich occurrences and is an iron ore in some localities.

Halides

Members of this class are distinguished by the large-sized anions of the halogens chlorine, bromine, iodine, and fluorine. The ions carry an electric charge of negative one and easily become distorted in the presence of strongly charged bodies. When associated with rather large, weakly polarizing cations of low charge, such as those of the alkali metals, both anions and cations take the form of nearly perfect spheres. Structures composed of these spheres exhibit the highest possible symmetry.

Pure ionic bonding is exemplified best in the isometric halides, for each spherical ion distributes its weak electrostatic charge over its entire surface. These halides manifest relatively low hardness and moderate-to-high melting points. In the solid state they are poor thermal and electric conductors, but when molten they conduct electricity well.

Halogen ions may also combine with smaller, more strongly polarizing cations than the alkali metal ions. Lower symmetry and a higher degree of covalent bonding prevail in these structures. Water and hydroxyl ions may enter the structure, as in atacamite $[Cu_2Cl(OH)_3]$.

The halides consist of about 80 chemically related minerals with diverse structures and widely varied origins. The most common are halite (NaCl), sylvite (KCl), chlorargyrite (AgCl), cryolite (Na_3AlF_6), fluorite (CaF_2), and atacamite. No molecules are present among the arrangement of the ions in halite, a naturally occurring form of sodium chloride. Each cation and anion is in octahedral coordination with its six closest neighbours. The NaCl structure is found in the crystals of many XZ-type halides, including sylvite (KCl) and chlorargyrite (AgCl). Some sulfides and oxides of XZ type crystallize in this structure type as well—for example, galena (PbS), alabandite (MnS), and periclase (MgO).

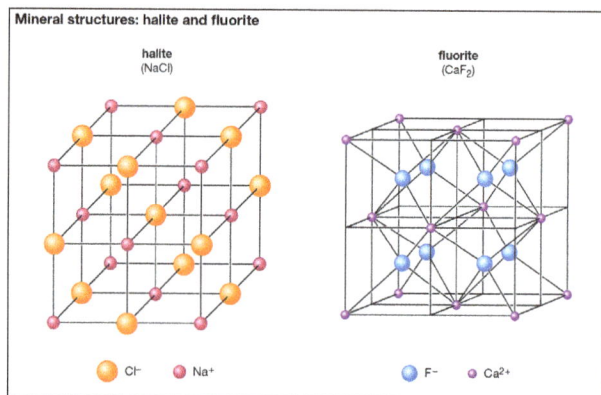

(A) The structure of halite, NaCl. (B) The structure of fluorite, CaF_2.

Several XZ_2 halides have the same structure as fluorite (CaF_2). In fluorite, calcium cations are positioned at the corners and face centres of cubic unit cells. (A unit cell is the smallest group of atoms, ions, or molecules from which the entire crystal structure can be generated by its repetition). Each fluorine anion is in tetrahedral coordination with four calcium ions, while each calcium cation is in eightfold coordination with eight fluorine ions that form the corners of a cube around it. Uraninite (UO_2) and thorianite (ThO_2) are two examples of the several oxides that have a fluorite-type structure.

Carbonates

The carbonate minerals contain the anionic complex $(CO_3)^{2-}$, which is triangular in its coordination—i.e., with a carbon atom at the centre and an oxygen atom at each of the corners of an equilateral triangle. These anionic groups are strongly bonded individual units and do not share oxygen atoms with one another. The triangular carbonate groups are the basic building units of all carbonate minerals and are largely responsible for the properties particular to the class.

Carbonates are frequently identified using the effervescence test with acid. The reaction that results in the characteristic fizz, $2H^+ + CO^2/_3 \rightarrow H_2O + CO_2$, makes use of the fact that the carbon-oxygen bonds of the CO_3 groups are not quite as strong as the corresponding carbon-oxygen bonds in carbon dioxide.

The common anhydrous (water-free) carbonates are divided into three groups that differ in structure type: calcite, aragonite, and dolomite. The copper carbonates azurite and malachite are the only notable hydrous varieties.

The members of the calcite group share a common structure type. It can be considered as a derivative of the NaCl structure in which the carbonate (CO_3) group substitute for the chlorine ions and calcium cations replace the sodium cations. As a result of the triangular shape of the CO_3 groups, the structure is rhombohedral instead of isometric as in NaCl. The CO_3 groups are in planes perpendicular to the threefold c-axis, and the calcium ions occupy alternate planes and are bonded to six oxygen atoms of the CO_3 groups.

Members of the calcite group exhibit perfect rhombohedral cleavage. The composition $CaCO_3$ most commonly occurs in two different polymorphs: rhombohedral calcite with calcium surrounded by six closest oxygen atoms and orthorhombic aragonite with calcium surrounded by nine closest oxygen atoms.

When CO_3 group are combined with large divalent cations (generally with ionic radii greater than 1.0 Å), orthorhombic structures result. This is known as the aragonite structure type. Members of this group include those with large cations: $BaCO_3$, $SrCO_3$, and $PbCO_3$. Each cation is surrounded by nine closest oxygen atoms.

The aragonite group displays more limited solid solution than the calcite group. The type of cation present in aragonite minerals is largely responsible for the differences in physical properties among the members of the group. Specific gravity, for example, is roughly proportional to the atomic weight of the metal ions.

Dolomite [$CaMg(CO_3)_2$], kutnohorite [$CaMn(CO_3)_2$], and ankerite [$CaFe(CO_3)_2$] are three isostructural members of the dolomite group. The dolomite structure can be considered as a calcite-type structure in which magnesium and calcium cations occupy the metal sites in alternate layers. The calcium (Ca^{2+}) and magnesium (Mg^{2+}) ions differ in size by 33 percent, and this produces cation ordering with the two cations occupying specific and separate levels in the structure. Dolomite has a calcium-to-magnesium ratio of approximately 1:1, which gives it a composition intermediate between $CaCO_3$ and $MgCO_3$.

Nitrates

The nitrates are characterized by their triangular $(NO_3)^-$ group that resemble the $(CO_3)^{2-}$ groups of the carbonates, making the two mineral classes similar in structure. The nitrogen cation (N^{5+}) carries a high charge and is strongly polarizing like the carbon cation (C^{4+}) of the CO_3 group. A tightly knit triangular complex is created by the three nitrogen-oxygen bonds of the NO_3 group; these bonds are stronger than all others in the crystal. Because the nitrogen-oxygen bond has greater strength than the corresponding carbon-oxygen bond in carbonates, nitrates decompose less readily in the presence of acids.

Nitrate structures analogous to those of the calcite group result when NO_3 combines in a 1:1 ratio with monovalent cations whose radii can accommodate six closest oxygen neighbours. For example, nitratite ($NaNO_3$) also called soda nitre, and calcite exhibit the same structure, crystallography, and cleavage. The two minerals differ in that nitratite is softer and melts at a lower temperature owing to its lesser charge; also, sodium has a lower atomic weight than calcium, causing nitratite to have a lower specific gravity as well. Similarly, nitre (KNO_3), also known as saltpetre, is an analogue of aragonite. These are two examples of only seven known naturally occurring nitrates.

Borates

Minerals of the borate class contain boron-oxygen groups that can link together, in a phenomenon known as polymerization, to form chains, sheets, and isolated multiple groups. The silicon-oxygen (SiO_4) tetrahedrons of the silicates polymerize in a manner similar to the $(BO_3)^{3-}$ triangular groups of the borates. A single oxygen atom is shared between two boron cations (B^{3+}), thereby linking the BO_3 groups into extended units such as double triangles, triple rings, sheets, and chains. The oxygen atom is able to accommodate two boron atoms because the small boron cation has a bond strength to each oxygen that is exactly one-half the bond energy of the oxygen ion.

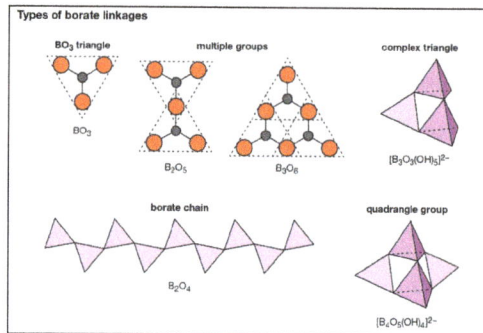

Various possible linkages of (A) BO_3 triangles to form (B,C) multiple groups and (D) chains in borates. Complex (E) triangle and (F) quadrangle groups are also shown. The group depicted in (F) occurs in borax.

Although boron is usually found in triangular coordination with three oxygens, it also occurs in four fold coordination in tetrahedral groups. In addition, boron may exist as part of complex anionic groups such as $[B_3O_3(OH)_3]^{2-}$, consisting of one triangle and two tetrahedrons. Complex infinite chains of tetrahedrons and triangles are found in the structure of colemanite $[CaB_3O_4(OH)_3 \cdot H_2O]$; a complex ion composed of two tetrahedrons and two triangles, $[B_4O_5(OH)_4]^{2-}$, is present in borax $[Na_2B_4O_5(OH)_4 \cdot 8H_2O]$.

Sulfates

This class is composed of a large number of minerals, but relatively few are common. All contain anionic $(SO_4)^{2-}$ groups in their structures. These anionic complexes are formed through the tight bonding of a central S^{6+} ion to four neighbouring oxygen atoms in a tetrahedral arrangement around the sulfur. This closely knit group is incapable of sharing any of its oxygen atoms with other SO_4 groups; as such, the tetrahedrons occur as individual, unlinked groups in sulfate mineral structures.

Common Sulfates	
Barite group	
Barite	$Baso_4$
Celestite	$Srso_4$

Anglesite	$PbSO_4$
Anhydrite	$CaSO_4$
Gypsum	$CaSO_4 \cdot 2H_2O$

Members of the barite group constitute the most important and common anhydrous sulfates. They have orthorhombic symmetry with large divalent cations bonded to the sulfate ion. In barite ($BaSO_4$), each barium ion is surrounded by 12 closest oxygen ions belonging to seven distinct SO_4 groups. Anhydrite ($CaSO_4$) exhibits a structure very different from that of barite since the ionic radius of Ca^{2+} is considerably smaller than Ba^{2+}. Each calcium cation can only fit eight oxygen atoms around it from neighbouring SO_4 groups. Gypsum ($CaSO_4 \cdot 2H_2O$) is the most important and abundant hydrous sulfate.

Phosphates

Although this mineral class is large (with almost 700 known species), most of its members are quite rare. Apatite [$Ca_5(PO_4)_3(F, Cl, OH)$], however, is one of the most important and abundant phosphates. The members of this group are characterized by tetrahedral anionic $(PO_4)^{3-}$ complexes, which are analogous to the $(SO_4)^{2-}$ groups of the sulfates. The phosphorus ion, with a valence of positive five, is only slightly larger than the sulfur ion, which carries a positive six charge. Arsenates and vanadates are similar to phosphates.

Silicates

The silicates, owing to their abundance on Earth, constitute the most important mineral class. Approximately 25 percent of all known minerals and 40 percent of the most common ones are silicates; the igneous rocks that make up more than 90 percent of Earth's crust are composed of virtually all silicates.

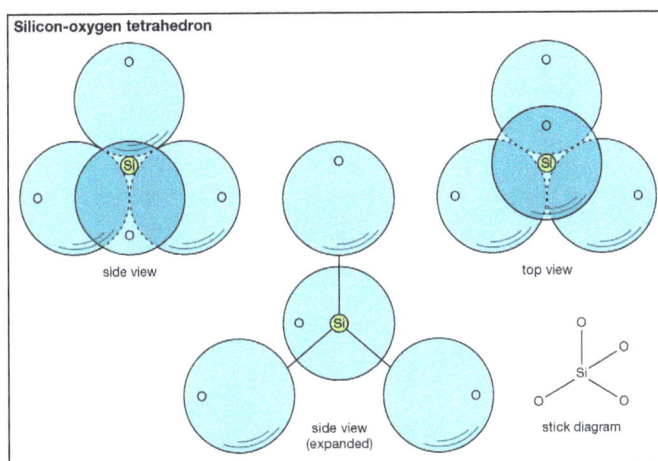

Silicon-oxygen tetrahedron

side view

top view

side view
(expanded)

stick diagram

Two views of a closest-packed representation of the silicon-oxygen tetrahedron.

The fundamental unit in all silicate structures is the silicon-oxygen (SiO_4)$^{4-}$ tetrahedron. It is composed of a central silicon cation (Si^{4+}) bonded to four oxygen atoms that are located at the corners of a regular tetrahedron. The terrestrial crust is held together by the strong silicon-oxygen bonds of these tetrahedrons. Approximately 50 percent ionic and 50 percent covalent, the bonds develop from the attraction of oppositely charged ions as well as the sharing of their electrons.

The positive charge (+4) of each silicon cation is satisfied by its four bonds to oxygen atoms. Each oxygen ion (O^{2-}), however, contributes only one-half of its total bonding energy to a silicon-oxygen bond, so it is capable of also bonding to the silicon cation of another tetrahedron. The SiO_4 tetrahedrons thereby become linked by shared oxygen atoms; this is referred to as polymerization. The degree and manner of polymerization are the bases for the variety present in silicate structures.

The silicates can be divided into groups according to structural configuration, which arises from the sharing of one, two, three, or all oxygen ions of a tetrahedron. Nesosilicates have isolated groups of SiO_4, while sorosilicates contain pairs of SiO_4 tetrahedrons linked into Si_2O_7 groups. Ring silicates, also known as cyclosilicates, are closed, ringlike silicates; the sixfold variety has composition Si_6O_{18}. Silicates that are composed of infinite chains of tetrahedrons are called inosilicates; single chains have a unit composition of SiO_3 or Si_2O_6, whereas double chains contain a silicon to oxygen ratio of 4:11. Phyllosilicates, or sheet silicates, are formed when three oxygen atoms are shared with adjoining tetrahedrons. The resulting infinite flat sheets have unit composition Si_2O_5. In structures where tetrahedrons share all their oxygen ions, an infinite three-dimensional network is created with an SiO_2 unit composition. Minerals of this type are called framework silicates or tectosilicates.

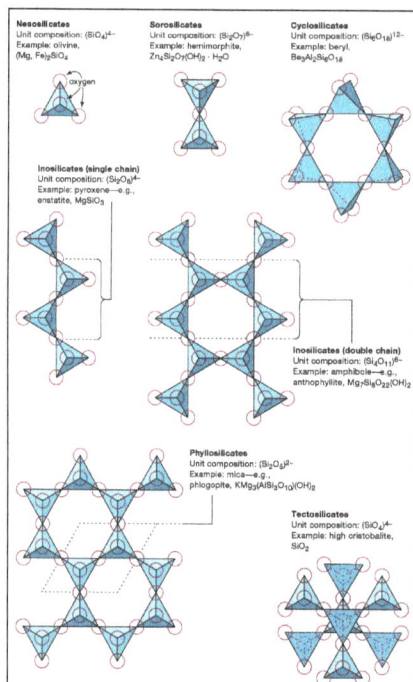

Figure: Various structural linkage schemes in silicates.

As a major constituent of Earth's crust, aluminum follows only oxygen and silicon in importance. The radius of aluminum, slightly larger than that of silicon, lies close to the upper bound for allowable fourfold coordination in crystals. As a result, aluminum can be surrounded with four oxygen atoms arranged tetrahedrally, but it can also occur in sixfold coordination with oxygen. The ability to maintain two roles within the silicate structure makes aluminum a unique constituent of these minerals. The tetrahedral AlO_4 groups are approximately equal in size to SiO_4 groups and therefore can become incorporated into the silicate polymerization scheme. Aluminum in sixfold coordination may form ionic bonds with the SiO_4 tetrahedrons. Thus, aluminum may occupy tetrahedral sites as a replacement for silicon and octahedral sites in solid solution with elements such as magnesium and ferrous iron.

Several ions may be present in silicate structures in octahedral coordination with oxygen: Mg^{2+}, Fe^{2+}, Fe^{3+}, Mn^{2+}, Al^{3+}, and Ti^{4+}. All cations have approximately the same dimensions and thus are found in equivalent atomic sites, even though their charges range from positive two to positive four. Solid solution involving ions of different charge is accomplished through coupled substitutions, thereby maintaining neutrality of the structures.

Nesosilicates

The silicon-oxygen tetrahedrons of the nesosilicates are not polymerized; they are linked to one another only by ionic bonds of the interstitial cations. As a result of the isolation of the tetrahedral groups, the crystal habits of these minerals are typically equidimensional so that prominent cleavage directions are not present. The size and charge of the interstitial cations largely determine the structural form of the nesosilicates. The relatively high specific gravity and hardness that are characteristic of this group arise from the dense packing of the atoms within the structure. Substitution of aluminum for silicon is normally quite low.

Sorosilicates

These minerals contain sets of two SiO_4 tetrahedrons joined by one shared apical oxygen. A silicon-to-oxygen ratio of 2:7 is consequently present in their structures. More than 70 minerals belong to the sorosilicate group, although most are rare. Only the members of the epidote group and vesuvianite are common. Both independent $(SiO_4)^{4-}$ and double $(Si_2O_7)^{6-}$ groups are incorporated into the epidote structure, as is reflected in its formula: $Ca_2(Al, Fe)Al_2O(SiO_4)(Si_2O_7)(OH)$.

Cyclosilicates

Silicon-oxygen tetrahedrons are linked into rings in cyclosilicate structures, which have an overall Si:O ratio of 1:3. There are three closed cyclic configurations with the following formulas: Si_3O_9, Si_4O_{12}, and Si_6O_{18}. The rare titanosilicate benitoite ($BaTiSi_3O_9$) is the only mineral that is built with the simple Si_3O_9 ring. Axinite [(Ca, Fe, Mn)$_3Al_2(BO_3)$ $(Si_4O_{12})(OH)$] contains Si_4O_{12} rings, along with BO_3 triangles and OH groups. The two common and important cyclosilicates, beryl (Be$_3Al_2Si_6O_{18}$) and tourmaline (which has an extremely complex formula), are based on the Si_6O_{18} ring.

Inosilicates

This class is characterized by its one-dimensional chains and bands created by the linkage of SiO$_4$ tetrahedrons. Single chains may be formed by the sharing of two oxygen atoms from each tetrahedron, resulting in a structure with an Si:O ratio of 1:3. Two such chains that are aligned side by side with alternate tetrahedrons sharing an additional oxygen atom form bands of double chains. These structures have an Si:O ratio of 4:11. There are a number of silicate minerals, pyroxenoids, which have a similar Si:O ratio as pyroxene, but with structures that are not identical as the chains of silicon tetrahedra do not infinitely repeat. Two significant rock-forming mineral families display these structure types: the single-chain pyroxenes and the double-chain amphiboles.

Inosilicates: Common pyroxenes and amphiboles	
Pyroxenes	
Enstatite-orthoferrosilite series	
Enstatite	Mgsio$_3$
Orthoferrosilite	Fesio$_3$
Diopside-hedenbergite series	
Diopside	Camgsi$_2$o$_6$
Hedenbergite	Cafesi$_2$o$_6$
Augite	(ca, na) (fe, mg, al) (al, si)$_2$o$_6$
Sodium pyroxene group	
Jadeite	Naalsi$_2$o$_6$
Acmite	Nafe^{3+}si$_2$o$_6$
Amphiboles	
Anthophyllite	(mg, fe)$_7$si$_8$o$_{22}$(oh)$_2$
Cummingtonite series	
Cummingtonite	Fe$_2$mg$_5$si$_8$o$_{22}$(oh)$_2$
Grunerite	Fe$_7$si$_8$o$_{22}$(oh)$_2$
Tremolite series	
Tremolite	Ca$_2$mg$_5$si$_8$o$_{22}$(oh)$_2$

Actinolite	$Ca_2(mg, fe)_5si_8O_{22}(oh)_2$
Hornblende	$(ca, na)_2(mg, fe, al)_5(si, al)_8O_{22}(oh)_2$
Sodic amphibole group	
Glaucophane	$Na_2mg_3al_2si_8O_{22}(oh)_2$
Riebeckite	$Na_2fe_3{}^{2+}fe_2{}^{3+}si_8O_{22}(oh)_2$

The amphiboles and pyroxenes share the same cations and have many similar crystallographic, chemical, and physical properties: the colour, lustre, and hardness of analogous species are alike. A distinguishing factor between the two groups, the presence of the hydroxyl radical in the amphiboles, generally gives the double-chain members lower specific gravities and refractive indices than their single-chain analogues. Their crystal habits also are different: amphiboles exhibit needle like or fibrous crystals, while pyroxenes take the form of stubby prisms. In addition, the different chain structures of the two groups result in different cleavage angles.

Pyroxenes occur in high-temperature igneous and metamorphic rocks. They crystallize at higher temperatures than their amphibole counterparts. A pyroxene formed early in the cooling of an igneous melt or in a metamorphic fluid may later combine with water at a lower temperature to form amphibole.

Phyllosilicates

These minerals display a two-dimensional framework of infinite sheets of SiO_4 tetrahedrons. An Si:O ratio of 2:5 results from the sharing of three oxygen atoms in each tetrahedron. Sixfold Symmetry is exhibited in undistorted sheets. The silicate sheet framework is largely responsible for the following properties of the phyllosilicates: platy or flaky habit, single pronounced cleavage, low specific gravity, softness, and possible flexibility and elasticity of cleavage layers. Most minerals of this group contain hydroxyls positioned in the middle of the sixfold rings of tetrahedrons.

Many soil constituents, produced through rock weathering, possess a sheet structure. Phyllosilicate properties contribute greatly to the ability of soils to release and retain plant food, to reserve water from wet to dry seasons, and to accommodate organisms and atmospheric gases.

Tectosilicates

Almost 75 percent of Earth's crust is composed of minerals with the three-dimensional framework of the tectosilicates. All oxygen atoms of the SiO_4 tetrahedrons of members of this class are shared with nearby tetrahedrons, creating a strongly bound structure with an Si:O ratio of 1:2. Other than the zeolite group, which can accommodate water owing to the open nature of its structure, all members listed in the table are anhydrous.

Rocks

In geology, Rock is naturally occurring and coherent aggregate of one or more minerals. Such aggregates constitute the basic unit of which the solid Earth is comprised and typically form recognizable and mappable volumes. Rocks are commonly divided into three major classes according to the processes that resulted in their formation. These classes are (1) igneous rocks, which have solidified from molten material called magma; (2) sedimentary rocks, those consisting of fragments derived from preexisting rocks or of materials precipitated from solutions; and (3) metamorphic rocks, which have been derived from either igneous or sedimentary rocks under conditions that caused changes in mineralogical composition, texture, and internal structure. These three classes, in turn, are subdivided into numerous groups and types on the basis of various factors, the most important of which are chemical, mineralogical, and textural attributes.

Physical Properties

Physical properties of rocks are of interest and utility in many fields of work, including geology, petrophysics, geophysics, materials science, geochemistry, and geotechnical engineering. The scale of investigation ranges from the molecular and crystalline up to terrestrial studies of the Earth and other planetary bodies. Geologists are interested in the radioactive age dating of rocks to reconstruct the origin of mineral deposits; seismologists formulate prospective earthquake predictions using premonitory physical or chemical changes; crystallographers study the synthesis of minerals with special optical or physical properties; exploration geophysicists investigate the variation of physical properties of subsurface rocks to make possible detection of natural resources such as oil and gas, geothermal energy, and ores of metals; geotechnical engineers examine the nature and behaviour of the materials on, in, or of which such structures as buildings, dams, tunnels, bridges, and underground storage vaults are to be constructed; solid-state physicists study the magnetic, electrical, and mechanical properties of materials for electronic devices, computer components, or high-performance ceramics; and petroleum reservoir engineers analyze the response measured on well logs or in the processes of deep drilling at elevated temperature and pressure.

Since rocks are aggregates of mineral grains or crystals, their properties are determined in large part by the properties of their various constituent minerals. In a rock these general properties are determined by averaging the relative properties and sometimes orientations of the various grains or crystals. As a result, some properties that are anisotropic (i.e., differ with direction) on a submicroscopic or crystalline scale are fairly isotropic for a large bulk volume of the rock. Many properties are also dependent on grain or crystal size, shape, and packing arrangement, the amount and distribution of void space, the presence of natural cements in sedimentary rocks, the temperature and pressure, and the type and amount of contained fluids (e.g., water, petroleum, gases).

Because many rocks exhibit a considerable range in these factors, the assignment of representative values for a particular property is often done using a statistical variation.

Some properties can vary considerably, depending on whether measured in situ (in place in the subsurface) or in the laboratory under simulated conditions. Electrical resistivity, for example, is highly dependent on the fluid content of the rock in situ and the temperature condition at the particular depth.

Density

Density varies significantly among different rock types because of differences in mineralogy and porosity. Knowledge of the distribution of underground rock densities can assist in interpreting subsurface geologic structure and rock type.

In strict usage, density is defined as the mass of a substance per unit volume; however, in common usage, it is taken to be the weight in air of a unit volume of a sample at a specific temperature. Weight is the force that gravitation exerts on a body (and thus varies with location), whereas mass (a measure of the matter in a body) is a fundamental property and is constant regardless of location. In routine density measurements of rocks, the sample weights are considered to be equivalent to their masses, because the discrepancy between weight and mass would result in less error on the computed density than the experimental errors introduced in the measurement of volume. Thus, density is often determined using weight rather than mass. Density should properly be reported in kilograms per cubic metre (kg/m^3), but is still often given in grams per cubic centimetre (g/cm^3).

Another property closely related to density is specific gravity. It is defined, as the ratio of the weight or mass in air of a unit volume of material at a stated temperature to the weight or mass in air of a unit volume of distilled water at the same temperature. Specific gravity is dimensionless (i.e., has no units).

The bulk density of a rock is $\rho B = W_G/V_B$, where W_G is the weight of grains (sedimentary rocks) or crystals (igneous and metamorphic rocks) and natural cements, if any, and V_B is the total volume of the grains or crystals plus the void (pore) space. The density can be dry if the pore space is empty, or it can be saturated if the pores are filled with fluid (e.g., water), which is more typical of the subsurface (in situ) situation. If there is pore fluid present,

$$(\text{saturated})\, \rho_B = \frac{W_G + W_{fl}}{V_B}$$

where, W_{fl} is the weight of pore fluid. In terms of total porosity, saturated density is

$$(\text{saturated})\, \rho_B = \rho_G (1 - \varphi r) + \rho fl \varphi r.$$

and thus $\varphi r = \dfrac{\rho_G - \rho_B}{\rho_G - \rho_{fl}}$,

where, ρ_{fl} is the density of the pore fluid. Density measurements for a given specimen involve the determination of any two of the following quantities: pore volume, bulk volume, or grain volume, along with the weight.

A useful way to assess the density of rocks is to make a histogram plot of the statistical range of a set of data. The representative value and its variation can be expressed as follows: (1) mean, the average value, (2) mode, the most common value (i.e., the peak of the distribution curve), (3) median, the value of the middle sample of the data set (i.e., the value at which half of the samples are below and half are above), and (4) standard deviation, a statistical measure of the spread of the data (plus and minus one standard deviation from the mean value includes about two-thirds of the data).

A compilation of dry bulk densities for various rock types found in the upper crust of the Earth is listed in the table. A histogram plot of these data, giving the percent of the samples as a function of density is shown in figure. The parameters given include (1) sample division, the range of density in one data column—e.g., 0.036 g/cm³ for figure below (2) number of samples, and (3) standard deviation. The small inset plot is the percentage of samples (on the vertical axis) that lie within the interval of the "mode - x" to the "mode + x," where x is the horizontal axis.

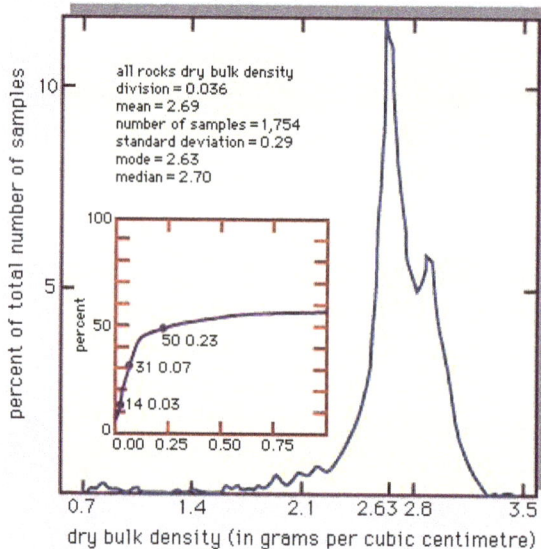

Figure: Dry bulk densities (distribution with density) for all rocks given in table.

In figure, the most common (modal) value of the distribution falls at 2.63 g/cm³, roughly the density of quartz, an abundant rock-forming mineral. Few density values for these upper crustal rocks lie above 3.3 g/cm³. A few fall well below the mode, even occasionally under 1 g/cm³. The reason for this is shown in figure above, which illustrates the density distributions for granite, basalt, and sandstone. Granite is an intrusive igneous rock with low porosity and a well-defined chemical (mineral) composition; its range of densities is narrow. Basalt is, in most cases, an extrusive igneous rock that can exhibit a large variation in porosity (because entrained gases leave voids called vesicles), and thus some highly porous samples can have low densities. Sandstone is a clastic sedimentary rock that can have a wide range of porosities depending on the degree of sorting, compaction, packing arrangement of grains, and cementation. The bulk density varies accordingly.

Other distribution plots of dry bulk densities are given in figures, with a sample division of 0.036 g/cm³ for figures and of 0.828 percent for figure.

The density of clastic sedimentary rocks increases as the rocks are progressively buried. This is because of the increase of overburden pressure, which causes compaction, and the progressive cementation with age. The bulk densities for sedimentary rocks, which typically have variable porosity, are given as ranges of both dry ρ_B and (water-) saturated ρ_B. The pore-filling fluid is usually briny water, often indicative of the presence of seawater when the rock was being deposited or lithified. It should be noted that the bulk density is less than the grain density of the constituent mineral (or mineral assemblage), depending on the porosity. For example, sandstone (characteristically quartzose) has a typical dry bulk density of 2.0–2.6 g/cm³, with a porosity that can vary from low to more than 30 percent. The density of quartz itself is 2.65 g/cm³. If porosity were zero, the bulk density would equal the grain density.

Saturated bulk density is higher than dry bulk density, owing to the added presence of pore-filling fluid.

Mechanical Properties

Stress and Strain

When a stress σ (force per unit area) is applied to a material such as rock, the material experiences a change in dimension, volume, or shape. This change, or deformation, is called strain (ε). Stresses can be axial—e.g., directional tension or simple compression—or shear (tangential), or all-sided (e.g., hydrostatic compression). The terms stress and pressure are sometimes used interchangeably, but often stress refers to directional stress or shear stress and pressure (P) refers to hydrostatic compression. For small stresses, the strain is elastic (recoverable when the stress is removed and linearly proportional to the applied stress). For larger stresses and other conditions, the strain can be inelastic, or permanent.

Elastic Constants

In elastic deformation, there are various constants that relate the magnitude of the strain response to the applied stress.

These elastic constants include the following:

(1) Young's modulus (E) is the ratio of the applied stress to the fractional extension (or shortening) of the sample length parallel to the tension (or compression). The strain is the linear change in dimension divided by the original length.

(2) Shear modulus (μ) is the ratio of the applied stress to the distortion (rotation) of a plane originally perpendicular to the applied shear stress; it is also termed the modulus of rigidity.

(3) Bulk modulus (k) is the ratio of the confining pressure to the fractional reduction of volume in response to the applied hydrostatic pressure. The volume strain is the change in volume of the sample divided by the original volume. Bulk modulus is also termed the modulus of incompressibility.

(4) Poisson's ratio (σ_p) is the ratio of lateral strain (perpendicular to an applied stress) to the longitudinal strain (parallel to applied stress).

For elastic and isotropic materials, the elastic constants are interrelated. For example,

$$\sigma_p = \frac{E}{2\mu} - 1;$$

$$E = 3k\left(1 - 2\sigma_p\right);$$

and

$$\mu = \frac{3}{2}k\left(\frac{1 - 2\sigma_p}{1 + \sigma_p}\right).$$

The following are the common units of stress:

1 bar = 10^6 dynes per square centimetre

= 10^5 newtons per square metre, or pascals (Pa)

= 0.1 megapascal (i.e., 0.1×10^6 pa)

Thus 10 kilobars = 1 gigapascal (i.e., 10^9 Pa).

Rock Mechanics

The study of deformation resulting from the strain of rocks in response to stresses is called rock mechanics. When the scale of the deformation is extended to large geologic structures in the crust of the Earth, the field of study is known as geotectonics.

The mechanisms and character of the deformation of rocks and Earth materials can be investigated through laboratory experiments, development of theoretical models based on the properties of materials, and study of deformed rocks and structures in the field. In the laboratory, one can simulate—either directly or by appropriate scaling of experimental parameters—several conditions. Two types of pressure may be simulated: confining (hydrostatic), due to burial under rock overburden, and internal (pore), due to pressure exerted by pore fluids contained in void space in the rock. Directed applied stress, such as compression, tension, and shear, is studied, as are the effects of increased temperature introduced with depth in the Earth's crust. The effects of the duration of time and the rate of applying stress (i.e., loading) as a function of time are examined. Also, the role of fluids, particularly if they are chemically active, is investigated.

Some simple apparatuses for deforming rocks are designed for biaxial stress application: a directed (uniaxial) compression is applied while a confining pressure is exerted (by pressurized fluid) around the cylindrical specimen. This simulates deformation at depth within the Earth. An independent internal pore-fluid pressure also can be exerted. The rock specimen can be jacketed with a thin, impermeable sleeve (e.g., rubber or copper) to separate the external pressure medium from the internal pore fluids. The specimen is typically a few centimetres in dimension.

Another apparatus for exerting high pressure on a sample was designed in 1968 by Akira Sawaoka, Naoto Kawai, and Robert Carmichael to give hydrostatic confining pressures up to 12 kilobars (1.2 gigapascal), additional directed stress, and temperatures up to a few hundred degrees Celsius. The specimen is positioned on the baseplate; the pressure is applied by driving in pistons with a hydraulic press. The end caps can be locked down to hold the pressure for time experiments and to make the device portable.

Apparatuses have been developed, typically using multianvil designs, which extend the range of static experimental conditions—at least for small specimens and limited times—to pressures as high as about 1,700 kilobars and temperatures of about 2,000° C. Such work has been pioneered by researchers such as Peter M. Bell and Ho-Kwang Mao, who conducted studies at the Geophysical Laboratory of the Carnegie Institution in Washington, D.C. Using dynamic techniques (i.e., shock from explosive impact generated by gun-type designs), even higher pressures up to 7,000 kilobars (700 gigapascal)—which is nearly twice the pressure at the centre of the Earth and seven million times greater than the atmospheric pressure at the Earth's surface—can be produced for very short times. A leading figure in such ultrapressure work is A. Sawaoka at the Tokyo Institute of Technology.

In the upper crust of the Earth, hydrostatic pressure increases at the rate of about 320 bars per kilometre, and temperature increases at a typical rate of 20°–40 °C per kilometre, depending on recent crustal geologic history. Additional directed stress, as can be generated by large-scale crustal deformation (tectonism), can range up to 1 to 2 kilobars. This is approximately equal to the ultimate strength (before fracture) of solid crystalline rock at surface temperature and pressure. The stress released in a single major earthquake—a shift on a fault plane—is about 50–150 bars.

In studying the deformation of rocks one can start with the assumption of ideal behaviour: elastic strain and homogeneous and isotropic stress and strain. In reality, on a microscopic scale there are grains and pores in sediments and a fabric of crystals in igneous and metamorphic rocks. On a large scale, rock bodies exhibit physical and chemical variations and structural features. Furthermore, conditions such as extended length of time, confining pressure, and subsurface fluids affect the rates of change of deformation. Figure shows the generalized transition from brittle fracture through faulting to plastic-flow deformation in response to applied compressional stress and the progressive increase of confining pressure.

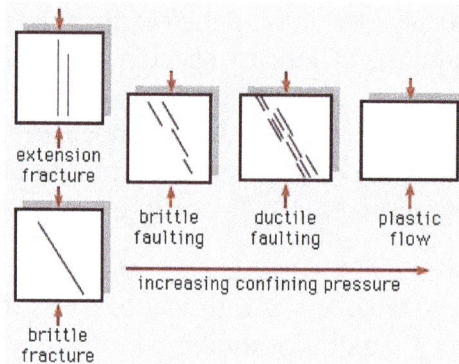

Figure: Deformation as affected by increased confining pressure.

Stress-strain Relationships

The deformation of materials is characterized by stress-strain relations. For elastic-behaviour materials, the strain is proportional to the load (i.e., the applied stress). The strain is immediate with stress and is reversible up to the yield point stress, beyond which permanent strain results. For viscous material, there is laminar (slow, smooth, parallel) flow; one must exert a force to maintain motion because of internal frictional resistance to flow, called the viscosity. Viscosity varies with the applied stress, strain rate, and temperature. In plastic behaviour, the material strains continuously after the yield point stress is reached; however, beyond this point there is some permanent deformation. In elasticoviscous deformation, there is combined elastic and viscous behaviour. The material yields continuously for a constant applied load. An example of such behaviour is creep, a slow, permanent, and continuous deformation occurring under constant load over a long time in such materials as crystals, ice, soil and sediment, and rocks at depth. In firmoviscous behaviour, the material is essentially solid but the strain is not immediate with application of stress; rather, it is taken up and released exponentially. A plasticoviscous material exhibits elastic behaviour for initial stress, but after the yield point stress is reached, it flows like a viscous fluid.

The coefficient of viscosity (η) is the ratio of applied stress to the rate of straining (change of strain with time). It is measured in units of poise; onepoise equals one dyne-second per square centimetre.

Rheology is the study of the flow deformation of materials. The concept of rheidity refers to the capacity of a material to flow, arbitrarily defined as the time required with a shear stress applied for the viscous strain to be 1,000 times greater than the elastic strain. It is thus a measure of the threshold of fluidlike behaviour. Although such behaviour depends on temperature, relative comparisons can be made.

Typical stress-strain (deformation) curves for rock materials are shown in figure. The stress σ, compression in the figure, is force per unit area. The strain ε is fractional shortening of the specimen parallel to the applied compression; it is given here in percent. The brittle material behaves elastically nearly until the point of fracture (denoted X), whereas the ductile (plastically deformable) material is elastic up to the yield point but then has a range of plastic deformation before fracturing. The ability to undergo large permanent deformation before fracture is called ductility. For plastic deformation, the flow mechanisms are intracrystalline (slip and twinning within crystal grains), intercrystalline motion by crushing and fracture (cataclasis), and recrystallization by solutioning or solid diffusion.

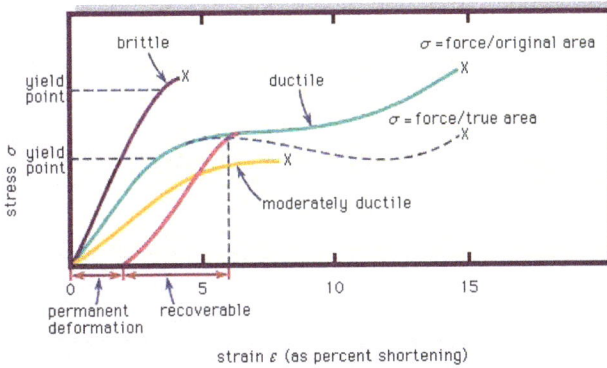

Figure: Typical stress-strain curves for rock materials. Each X represents
the point of fracture for the corresponding material.

If the applied stress is removed while a ductile material is in the plastic range, part of the strain is recoverable, but there is permanent deformation. The ultimate strength is the highest point on a stress-strain curve, often occurring at fracture (which is the complete loss of cohesion). The strength of a material is its resistance to failure by flow or fracture; it is a measure of the stress required to deform a body.

Effect of Environmental Conditions

The behaviour and mechanical properties of rocks depend on a number of environmental conditions. (1) Confining pressure increases the elasticity, strength (e.g., yield point and ultimate fracture stress), and ductility. (2) Internal pore-fluid pressure reduces the effective stress acting on the sample, thus reducing the strength and ductility. The effective, or net, confining pressure is the external hydrostatic pressure minus the internal pore-fluid pressure. (3) Temperature lowers the strength, enhances ductility, and may enhance recrystallization. (4) Fluid solutions can enhance deformation, creep, and recrystallization. (5) Time is an influential factor as well. (6) The rate of loading (i.e., the rate at which stress is applied) influences mechanical properties. (7) Compaction, as would occur with burial to depth, reduces the volume of pore space for sedimentary rocks and the crack porosity for crystalline rocks.

Rocks, which are typically brittle at the Earth's surface, can undergo ductile deformation when buried and subjected to increased confining pressure and temperature for long periods of time. If stress exceeds their strength or if they are not sufficiently ductile, they will fail by fracture—as a crystal, within a bed or rock, on an earthquake fault zone, and so on—whereas with ductility they can flow and fold.

The plastic yield strength here is the stress at a 2 percent strain; the ultimate strength, as stated above, is the highest point on the stress-strain curve.

An increase in confining pressure causes brittle fracture to become shear slippage and eventually causes flow behaviour. This transition is also aided by higher temperature, decreased internal pore-fluid pressure, and slower strain rate.

Thermal Properties

Heat flow (or flux), q, in the Earth's crust or in rock as a building material, is the product of the temperature gradient (change in temperature per unit distance) and the material's thermal conductivity (k, the heat flow across a surface per unit area per unit time when a temperature difference exists in unit length perpendicular to the surface). Thus,

$$q = \frac{dT}{dz} \times k.$$

The units of the terms in this equation are given below, expressed first in the centimetre-gram-second (cgs) system and then in the International System of Units (SI) system, with the conversion factor from the first to the second given between them.

q, heat flow:
    calories per square
    centimetre per second $\times \dfrac{1}{23.9} \times 10^{-6} =$ watts per
    square metre
    $(cal / cm^2 . sec)$ (W/m²).

$\dfrac{dT}{dz}$, temperature gradient: $\times 10^{-3} =$ degree Celsius
    Degree Celsius per   per metre
    Kilometre (°C/Km,   (°C/m).
    Practical unit for Earth)

K, thermal conductivity:
    Calories per square
    Centimetre per second
    Per degree Celsius
    $(cal/cm^2 \cdot sec \cdot {}^{\circ}C)$

$$\times \frac{1}{23.9} \times 10^{-3} =$$

watts per
metre per
degree Celsius
$(W/m \cdot {}^{\circ}C)$

Thermal Conductivity

Thermal conductivity can be determined in the laboratory or in situ, as in a borehole or deep well, by turning on a heating element and measuring the rise in temperature with time. It depends on several factors: (1) chemical composition of the rock (i.e., mineral content), (2) fluid content (type and degree of saturation of the pore space); the presence of water increases the thermal conductivity (i.e., enhances the flow of heat), (3) pressure (a high pressure increases the thermal conductivity by closing cracks which inhibit heat flow), (4) temperature, and (5) isotropy and homogeneity of the rock.

Typical values of thermal conductivities of rock materials are given in the table. For crystalline silicate rocks—the dominant rocks of the "basement" crustal rocks—the lower values are typical of ones rich in magnesium and iron (e.g., basalt and gabbro) and the higher values are typical of those rich in silica (quartz) and alumina (e.g., granite). These values result because the thermal conductivity of quartz is relatively high, while that for feldspars is low.

Thermal Expansion

The change in dimension—linear or volumetric—of a rock specimen with temperature is expressed in terms of a coefficient of thermal expansion. This is given as the ratio of dimension change (e.g., change in volume) to the original dimension (volume, V) per unit of temperature (T) change:

$$\frac{1}{V} = \frac{\Delta V}{\Delta T}.$$

Most rocks have a volume-expansion coefficient in the range of $15^{-33} \times 10^{-6}$ per degree Celsius under ordinary conditions. Quartz-rich rocks have relatively high values because of the higher volume expansion coefficient of quartz. Thermal-expansion coefficients increase with temperature.

$$\frac{1}{L} = \frac{\Delta L}{\Delta T}.$$

Where L represents length. All data are based on at least three samples.

Radioactive Heat Generation

The spontaneous decay (partial disintegration) of the nuclei of radioactive elements provides decay particles and energy. The energy, composed of emission kinetic energy and radiation, is converted to heat; it has been an important factor in affecting the temperature gradient and thermal evolution of the Earth. Deep-seated elevated temperatures provide the heat that causes rock to deform plastically and to move, thus generating to a large extent the processes of plate tectonics—plate motions, seafloor spreading, continental drift, and subduction—and most earthquakes and volcanism.

Some elements, or their isotopes (nuclear species with the same atomic number but different mass numbers), decay with time. These include elements with an atomic number greater than 83—of which the most important are uranium-235, uranium-238, and thorium-232—and a few with a lower atomic number, such as potassium-40.

The heat generated within rocks depends on the types and abundances of the radioactive elements and their host minerals. Such heat production, A, is given in calories per cubic centimetre per second, or 1 calorie per gram per year = 4.186×10^7 ergs per gram per year = 1.327 ergs per gram per second. The rate of radioactive decay, statistically an exponential process, is given by the half-life, t1/2. The half-life is the time required for half the original radioactive atoms to decay for a particular isotope.

The isotopic abundance is the percent of the natural element that exists as that particular radioactive isotope; for example, 99.28 percent of natural uranium is U-238, and 100 percent of thorium is the radioactive Th-232. The final product is the end result of the process (usually multistage) of disintegration. For the rocks, the typical content is given for uranium and thorium (in parts per million [ppm] of weight) and for potassium (in weight percent). The heat production of natural uranium is close to that for the isotope U-238, since almost all natural uranium is of that isotopic species.

The radioactive elements are more concentrated in the continental upper-crust rocks that are rich in quartz (i.e., felsic, or less mafic). This results because these rocks are differentiated by partial melting of the upper-mantle and oceanic-crust rock. The radioactive elements tend to be preferentially driven off from these rocks for geochemical reasons.

Electrical Properties

The electrical nature of a material is characterized by its conductivity (or, inversely, its resistivity) and its dielectric constant, and coefficients that indicate the rates of change of these with temperature, frequency at which measurement is made, and so on. For rocks with a range of chemical composition as well as variable physical properties of porosity and fluid content, the values of electrical properties can vary widely.

Resistance (R) is defined as being one ohm when a potential difference (voltage; V) across a specimen of one volt magnitude produces a current (i) of one ampere; that is, V = Ri. The electrical resistivity (ρ) is an intrinsic property of the material. In other words, it is inherent and not dependent on sample size or current path. It is related to resistance by R = ρL/A where L is the length of specimen, A is the cross-sectional area of specimen, and units of ρ are ohm-centimetre; 1 ohm-centimetre equals 0.01 ohm-metre. The conductivity (σ) is equal to $1/\rho$ ohm $^{-1} \cdot$ centimetre^{-1} (or termed mhos/cm). In SI units, it is given in mhos/metre, or siemens/metre.

Some representative values of electrical resistivity for rocks and other materials. Materials that are generally considered as "good" conductors have a resistivity of 10^{-5}–10 ohm-centimetre (10^{-7}–10^{-1} ohm-metre) and a conductivity of 10–10^7 mhos/metre. Those that are classified as intermediate conductors have a resistivity of 100–10^9 ohm-centimetre (1–10^7 ohm-metre) and a conductivity of 10^{-7}–1 mhos/metre. "Poor" conductors, also known as insulators, have a resistivity of 10^{10}–10^{17} ohm-centimetre (10^8–10^{15} ohm-metre) and a conductivity of 10^{-15}–10^{-8}. Seawater is a much better conductor (i.e., it has lower resistivity) than fresh water owing to its higher content of dissolved salts; dry rock is very resistive. In the subsurface, pores are typically filled to some degree by fluids. The resistivity of materials has a wide range—copper is, for example, different from quartz by 22 orders of magnitude.

For high-frequency alternating currents, the electrical response of a rock is governed in part by the dielectric constant, ε. This is the capacity of the rock to store electric charge; it is a measure of polarizability in an electric field. In cgs units, the dielectric constant is 1.0 in a vacuum. In SI units, it is given in farads per metre or in terms of the ratio of specific capacity of the material to specific capacity of vacuum (which is 8.85×10^{-12} farads per metre). The dielectric constant is a function of temperature, and of frequency, for those frequencies well above 100 hertz (cycles per second).

Electrical conduction occurs in rocks by (1) fluid conduction—i.e., electrolytic conduction by ionic transfer in briny pore water—and (2) metallic and semiconductor (e.g., some sulfide ores) electron conduction. If the rock has any porosity and contained fluid, the fluid typically dominates the conductivity response. The rock conductivity depends on the conductivity of the fluid (and its chemical composition), degree of fluid saturation, porosity and permeability, and temperature. If rocks lose water, as with compaction of clastic sedimentary rocks at depth, their resistivity typically increases.

Magnetic Properties

The magnetic properties of rocks arise from the magnetic properties of the constituent mineral grains and crystals. Typically, only a small fraction of the rock consists of magnetic minerals. It is this small portion of grains that determines the magnetic properties and magnetization of the rock as a whole, with two results: (1) the magnetic properties of a given rock may vary widely within a given rock body or structure, depending on chemical inhomogeneities, depositional or crystallization conditions, and what happens to the rock after formation; and (2) rocks that share the same lithology (type and name) need not necessarily share the same magnetic characteristics. Lithologic classifications are usually based on the abundance of dominant silicate minerals, but the magnetization is determined by the minor fraction of such magnetic mineral grains as iron oxides. The major rock-forming magnetic minerals are iron oxides and sulfides.

Although the magnetic properties of rocks sharing the same classification may vary from rock to rock, general magnetic properties do nonetheless usually depend on rock type and overall composition. The magnetic properties of a particular rock can be quite well understood provided one has specific information about the magnetic properties of crystalline materials and minerals, as well as about how those properties are affected by such factors as temperature, pressure, chemical composition, and the size of the grains. Understanding is further enhanced by information about how the properties of typical rocks are dependent on the geologic environment and how they vary with different conditions.

Applications of the Study of Rock Magnetization

An understanding of rock magnetization is important in at least three different areas: prospecting, geology, and materials science. In magnetic prospecting, one is interested in mapping the depth, size, type, and inferred composition of buried rocks. The prospecting, which may be done from ground surface, ship, or aircraft, provides an important first step in exploring buried geologic structures and may, for example, help identify favourable locations for oil, natural gas, and economic mineral deposits.

Rock magnetization has traditionally played an important role in geology. Paleomagnetic work seeks to determine the remanent magnetization and thereby ascertain the character of the Earth's field when certain rocks were formed. The results of such research have important ramifications in stratigraphic correlation, age dating, and reconstructing past movements of the Earth's crust. Indeed, magnetic surveys of the oceanic crust provided for the first time the quantitative evidence needed to cogently demonstrate that segments of the crust had undergone large-scale lateral displacements over geologic time, thereby corroborating the concepts of continental drift and seafloor spreading, both of which are fundamental to the theory of plate tectonics .

The understanding of magnetization is increasingly important in materials science as well. The design and manufacture of efficient memory cores, magnetic tapes, and permanent magnets increasingly rely on the ability to create materials having desired magnetic properties.

Basic Types of Magnetization

There are six basic types of magnetization: (1) diamagnetism, (2) paramagnetism, (3) ferromagnetism, (4) antiferromagnetism, (5) ferrimagnetism, and (6) superparamagnetism.

Diamagnetism arises from the orbiting electrons surrounding each atomic nucleus. When an external magnetic field is applied, the orbits are shifted in such a way that the atoms set up their own magnetic field in opposition to the applied field. In other words, the induced diamagnetic field opposes the external field. Diamagnetism is present in all materials, is weak, and exists only in the presence of an applied field. The propensity of a substance for being magnetized in an external field is called its susceptibility (k) and it is defined as J/H, where J is the magnetization (intensity) per unit volume and H is the strength of the applied field. Since the induced field always opposes the applied field, the sign of diamagnetic susceptibility is negative. The susceptibility of a diamagnetic substance is on the order of -10^{-6} electromagnetic units per cubic centimetre (emu/cm^3). It is sometimes denoted κ for susceptibility per unit mass of material.

Paramagnetism results from the electron spin of unpaired electrons. An electron has a magnetic dipole moment—which is to say that it behaves like a tiny bar magnet—and so when a group of electrons is placed in a magnetic field, the dipole moments tend to line up with the field. The effect augments the net magnetization in the direction of the applied field. Like diamagnetism, paramagnetism is weak and exists only in the presence of an applied field, but since the effect enhances the applied field, the sign of the paramagnetic susceptibility is always positive. The susceptibility of a paramagnetic substance is on the order of 10^{-4} to 10^{-6} emu/cm^3.

Ferromagnetism also exists because of the magnetic properties of the electron. Unlike paramagnetism, however, ferromagnetism can occur even if no external field is applied. The magnetic dipole moments of the atoms spontaneously line up with one another because it is energetically favourable for them to do so. A remanent magnetization can be retained. Complete alignment of the dipole moments would take place only at a temperature of absolute zero (0 kelvin [K], or -273.15° C). Above absolute zero, thermal motions begin to disorder the magnetic moments. At a temperature called the Curie temperature, which varies from material to material, the thermally induced disorder overcomes the alignment, and the ferromagnetic properties of the substance disappear. The susceptibility of ferromagnetic materials is large and positive. It is on the order of 10 to 10^4 emu/cm^3. Only a few materials—iron, cobalt, and nickel—are ferromagnetic in the strict sense of the word and have a strong residual magnetization. In general usage, particularly in engineering, the term ferromagnetic is frequently applied to any material that is appreciably magnetic.

Antiferromagnetism occurs when the dipole moments of the atoms in a material assume an antiparallel arrangement in the absence of an applied field. The result is that the sample has no net magnetization. The strength of the susceptibility is comparable to that of paramagnetic materials. Above a temperature called the Néel temperature, thermal motions destroy the antiparallel arrangement, and the material then becomes paramagnetic. Spin-canted (anti)ferromagnetism is a special condition which occurs when antiparallel magnetic moments are deflected from the antiferromagnetic plane, resulting in a weak net magnetism. Hematite (α-Fe$_{2O_3}$) is such a material.

Ferrimagnetism is an antiparallel alignment of atomic dipole moments which does yield an appreciable net magnetization resulting from unequal moments of the magnetic sublattices. Remanent magnetization is detectable. Above the Curie temperature the substance becomes paramagnetic. Magnetite (Fe_3O_4), which is the most magnetic common mineral, is a ferrimagnetic substance.

Superpara magnetism occurs in materials having grains so small (about 100 angstroms) that any cooperative alignment of dipole moments is overcome by thermal energy.

Types of Remanent Magnetization

Rocks and minerals may retain magnetization after the removal of an externally applied field, thereby becoming permanent weak magnets. This property is known as remanent magnetization and is manifested in different forms, depending on the magnetic properties of the rocks and minerals and their geologic origin and history. Delineated below are the kinds of remanent magnetization frequently observed:

- CRM (chemical, or crystallization, remanent magnetization) can be induced after a crystal is formed and undergoes one of a number of physicochemical changes, such as oxidation or reduction, a phase change, dehydration, recrystallization, or precipitation of natural cements. The induction, which is particularly important in some (red) sediments and metamorphic rocks, typically takes place at constant temperature in the Earth's magnetic field.

- DRM (depositional, or detrital, remanent magnetization) is formed in clastic sediments when fine particles are deposited on the floor of a body of water. Marine sediments, lake sediments, and some clays can acquire DRM. The Earth's magnetic field aligns the grains, yielding a preferred direction of magnetization.

- IRM (isothermal remanent magnetization) results from the application of a magnetic field at a constant (isothermal) temperature, often room temperature.

- NRM (natural remanent magnetization) is the magnetization detected in a geologic in situ condition. The NRM of a substance may, of course, be a combination of any of the other remanent magnetizations.

- PRM (pressure remanent, or piezoremanent, magnetization) arises when a material undergoes mechanical deformation while in a magnetic field. The process of deformation may result from hydrostatic pressure, shock impact (as produced by a meteorite striking the Earth's surface), or directed tectonic stress. There are magnetization changes with stress in the elastic range, but the most pronounced effects occur with plastic deformation when the structure of the magnetic minerals is irreversibly changed.

- TRM (thermoremanent magnetization) occurs when a substance is cooled, in the presence of a magnetic field, from above its Curie temperature to below that temperature. This form of magnetization is generally the most important, because it is stable and widespread, occurring in igneous and sedimentary rocks. TRM also can occur when dealing exclusively with temperatures below the Curie temperature. In PTRM (partial thermoremanent magnetization) a sample is cooled from a temperature below the Curie point to yet a lower temperature.

- VRM (viscous remanent magnetization) results from thermal agitation. It is acquired slowly over time at low temperatures and in the Earth's magnetic field. The effect is weak and unstable but is present in most rocks.

Hysteresis and Magnetic Susceptibility

The concept of hysteresis is fundamental when describing and comparing the magnetic properties of rocks. Hysteresis is the variation of magnetization with applied field and illustrates the ability of a material to retain its magnetization, even after an applied field is removed. Figure illustrates this phenomenon in the form of a plot of magnetization (J) versus applied field (H_{ex}). J_s is the saturation (or "spontaneous") magnetization when all the magnetic moments are aligned in their configuration of maximum order. It is temperature-dependent, reaching zero at the Curie temperature. Jr sat is the remanent magnetization that remains when a saturating (large) applied field is removed, and Jr is the residual magnetization left by some process apart from IRM saturation, as, for example, TRM. $H_{c,r}$ is the coercive field (or force) that is required to reduce $J_{r,sat}$ to zero, and $H_{c,r}$ is the field required to reduce J_r to zero.

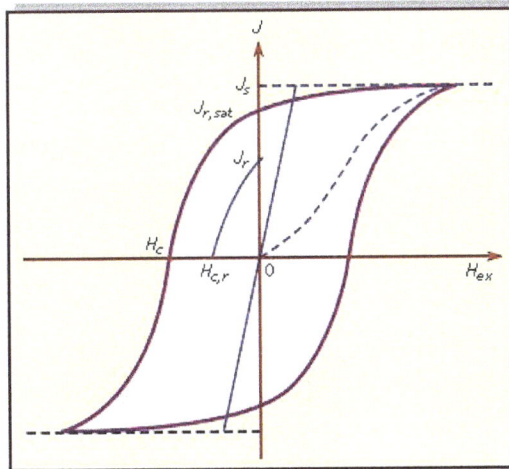

The above figure mentions the General magnetic hysteresis curve, showing magnetization (J) as a function of the external field (Hex). J_s is the saturation (or "spontaneous") magnetization; $J_{r,sat}$ is the remanent magnetization that remains after a saturating applied field is removed; J_r is the residual magnetization left by some magnetization process other than IRM saturation; H_c is the coercive field; and $H_{c,r}$ is the field necessary to reduce J_r to zero.

Magnetic susceptibility is a parameter of considerable diagnostic and interpretational use in the study of rocks. This is true whether an investigation is being conducted in the laboratory or magnetic fields over a terrain are being studied to deduce the structure and lithologic character of buried rock bodies. Susceptibility for a rock type can vary widely, depending on magnetic mineralogy, grain size and shape, and the relative magnitude of remanent magnetization present, in addition to the induced magnetization from the Earth's weak field. The latter is given as $J_{induced} = k_{Hex}$, where k is the (true) magnetic susceptibility and H_{ex} is the external (i.e., the Earth's) magnetic field. If there is an additional remanent magnetization with its ratio (Q_n) to induced magnetization being given by:

$$Q_n = \frac{J_{remanent}}{kH_{ex}},$$

then the total magnetization is:

$$J = J_{induced} + J_{remanent}$$
$$= kH_{ex} + Q_n kH_{ex}$$
$$= k\left(1 + Q_n\right)H_{ex}$$
$$= K_{app}H_{ex}.$$

Where kapp, the "apparent" magnetic susceptibility, is $k(1 + Q_n)$.

Magnetic Minerals and Magnetic Properties of Rocks

The major rock-forming magnetic minerals are the following iron oxides: the titano-magnetite series, $xFe_2TiO_4 \cdot (1 - x)Fe_3O_4$, where Fe_3O_4 is magnetite, the most magnetic mineral; the ilmenohematite series, $yFeTiO_3 \cdot (1 - y)Fe_2O_3$, where $\alpha\text{-}Fe_2O_3$ (in its rhombohedral structure) is hematite; maghemite, $\gamma\text{-}Fe_2O_3$ (in which some iron atoms are missing in the hematite structure); and limonite (hydrous iron oxides). They also include sulfides—namely, the pyrrhotite series, $yFeS \cdot (1 - y)Fe_{1-x}S$.

A distribution of measured (true) susceptibilities for various rock types is shown in the table. Basic refers to those rocks high in iron and magnesium silicates and magnetite, extrusive means formed by cooling after extruding onto the land surface or seafloor. The data in each category are based on at least 45 samples. Natural remanent magnetization is some combination of remanences; typically TRM in an igneous rock, perhaps DRM or CRM or both in a sedimentary rock, and all with an additional VRM. The ratio Q_n is typically higher for rocks with a strong, stable remanence—e.g., magnetite-rich and fine-grained extrusive rocks such as seafloor basalts.

Rock Types

The three types of rocks are:

- Igneous: They form from the cooling of magma deep inside the earth. They often have large crystals.

- Metamorphic: They are formed through the change (metamorphosis) of igneous and sedimentary rocks. They can form both underground and at the surface.

- Sedimentary: They are formed through the solidification of sediment. They can be formed from organic remains (such as limestone), or from the cementing of other rocks.

Igneous Rocks

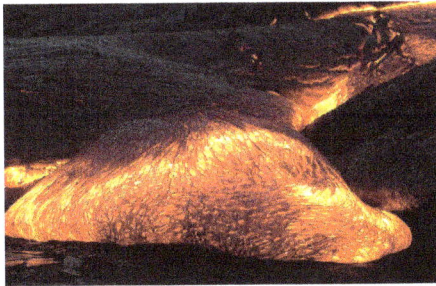

Lava flow on Hawaii-Lava is the extrusive equivalent of magma.

Magma is the heart of any igneous rock. Magma is composed of a mixture of molten or semi-molten rock, along with gases and other volatile elements. As you go deeper underground, the temperature rises; go further and you'll eventually reach the Earth's mantle —a huge layer of magma surrounding the Earth's core.

As you probably know, when magma cools, it turns into rock; if it cools while still underground at high temperatures (but at temperatures still lower than that of the magma), the cooling process will be slow, giving crystals time to develop. That's why you see rocks such as granite with big crystals — the magma had time to cool off.

Note the white, almost rectangular feldspar crystals, the grey virtually shapeless quartz crystals, and the black crystals, which can be either black mica or amphibole.

However, if the magma erupts or is cooled rapidly, you instead get a volcanic rock – not really igneous, but also originating from lava. The classical example here is basalt, which can have many small crystals or very few large ones. Volcanic rocks are also called *extrusive igneous* rocks, as opposed to *intrusive igneous* rocks. Some volcanic rocks (like obsidian) don't have any crystals at all.

Basalt

Pumice

Not all magma is made equally: different magmas can have different chemical compositions, different quantities of gases and different temperature and different types of magma make different types of rocks. That's why you get incredible variety. There are over 700 hundred types of igneous rocks, and they are generally the hardest and heaviest of all rocks. However, volcanic rocks can be incredibly lightweight–pumice, for example, can even float, and was called by ancient sailors "the foam of the sea". Pumice is created when a volcano violently erupts, creating pockets of air in the rock. The most common types of igneous rocks are:

- Andesite

- Basalt

- Dacite

- Dolerite (also called diabase)

- Gabbro

- Diorite

- Peridotite

- Nepheline

- Obsidian
- Scoria

- Tuff
- Volcanic bomb.

Metamorphic Rocks

Here, the name says it all. These are rocks that underwent a metamorphosis; they changed. They were either sedimentary or igneous (or even metamorphic), and they changed so much, that they are fundamentally different from the initial rock.

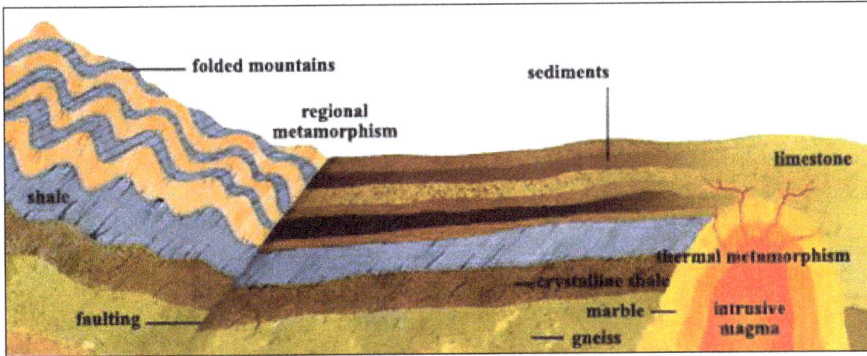

Different types of metamorphism.

There are two types of metamorphism (change) that can cause this:

- Contact metamorphism (or thermal metamorphism) — rocks are so close to magma that they start to partially melt and change their properties. You can have recrystallization, fusing between crystals and a lot of other chemical reactions. Temperature is the main driver here.

- Regional metamorphism (or dynamic metamorphism) — this typically happens when rocks are deep underground and they are subjected to massive pressure — so much so that they often become elongated, destroying the original features. Pressure (often times with temperature) is the main driver here.

Folded foliation in a metamorphic rock from near Geirangerfjord, Norway.

Metamorphic rocks can have crystals and minerals from the initial rocks as well as new minerals resulting from the metamorphosis process. However, some minerals are clear indicators of a metamorphic process. Among these, the most usual ones are garnet, chlorite, and kyanite.

Equally as significant are changes in the chemical environment that result in two metamorphic processes: mechanical dislocation (the rock or some minerals are physically altered) and chemical recrystallization (when the temperature and pressure changes, some crystals aren't stable, causing them to change into other crystals).

Marble is a non-foliated metamorphic rock.

They can be divided into many categories, but they are typically split into:

- Foliated metamorphic rocks: Pressure squeezes or elongates the crystals, resulting in a clear preferential alignment.

- Non-foliated metamorphic rocks: The crystals have no preferential alignment. Some rocks, such as limestone, are made of minerals that simply don't elongate, no matter how much stress you apply.

Metamorphic rocks can form in different conditions, in different temperatures (up to 200 °C) and pressures (up to 1500 bars). By being buried deep enough for a long enough time, a rock will become metamorphic. They can form from tectonic processes such as continental collisions, which cause horizontal pressure, friction and distortion; they can also form when the rock is heated up by the intrusion of magma from the Earth's interior.

The most common metamorphic rocks are:

- Amphibolite

- Schist (blueschist, greenschist, micaschist, etc.)

- Eclogite

- Gneiss

- Hornfels

- Marble

- Migmatite

- Phyllite

- Quartzite

- Serpentinite

- Slate.

A micaschist: The dark brown rounded minerals are garnet, and everything you see with a whiteish tint is the mica. The reddish areas are rusty mica.

Sedimentary Rocks

Sedimentary rocks are named as such because they were once sediment. Sediment is a naturally occurring material that is broken down by the processes of weathering and erosion and is subsequently naturally transported (or not). Sedimentary rocks form through the deposition of material at the Earth's surface and within bodies of water.

A conglomerate — a rock made from cemented gravel.

Sedimentary rocks are quite difficult to classify, as they have several different defining qualities (the chemical make-up, the sedimentation process, organic/inorganic material), but the most common classification is the following:

- Clastic sedimentary rocks — small rock fragments (many silicates) that were transported and deposited by fluids (water, bed flows). These rocks are further classified by the size and composition of the clastic crystals included in the sedimentary rocks (most often quartz, feldspar, mica and clay).

- Conglomerates (and breccias) — conglomerates are predominantly composed of rounded gravel, while breccias are composed of angular (sharper) gravel.

- Sandstones — as the name says, it's a rock made from many-sand-sized minerals and rock grains. The most dominant mineral in sandstone is quartz because it is the most common mineral in the Earth's surface crust.

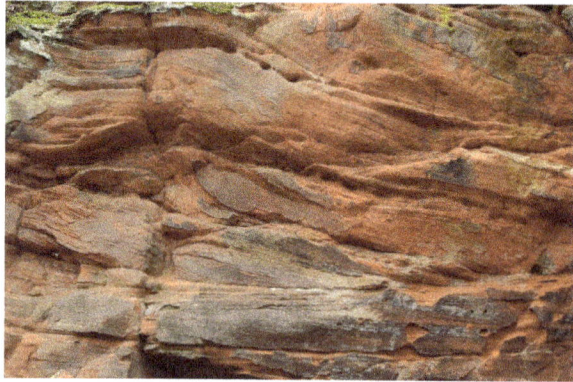

An old, red sandstone.

- Mudrocks — again, the name says it all — they're rocks made from solidified mud. They typically contain very fine particles and are transported as suspended particles by turbulent flow in water or air, depositing once the flow settles.

- Biochemical rocks — you'll probably be surprised to find out that most limestone on the face of the earth comes from biological sources. In other words, most limestone you see today comes from the skeletons of organisms such as corals, mollusks, and foraminifera. Coal is another example of biochemical rock.

- Chemical rocks — these rocks include gypsum and salt (halite) and are formed mostly through water evaporation.

There are also other types of specific sedimentary rocks — for example, the ones formed in hot springs. Most of the solid surface of our planet (roughly 70%) is represented by sedimentary rocks, but if you go deep enough beneath the Earth's surface, there are plenty of igneous and metamorphic rocks to be found.

Yes, salt is a mineral — and it can be quite beautiful. In this context, it's called halite and can be classified as a sedimentary rock.

This entire mountain in Romania was formed based on a coral reef.

Some common sedimentary rocks are:

- Argillite

- Breccia

- Chalk

- Chert

- Claystone

- Coal

- Conglomerate

- Dolomite

- Limestone

- Gypsum

- Greywacke

- Mudstone

- Shale

- Siltstone

- Turbidite.

Soil

Soil is Earth's element. It is the mixture of minerals, organic matter, liquids, and gases covering most of the Earth's land surface and that serves, or has the ability to serve, as a medium for the growth of land plants. Although it may be covered by shallow water, if the water is too deep to support land plants (typically more than 2.5 meters), then the rock-covering mixture is not considered to be soil.

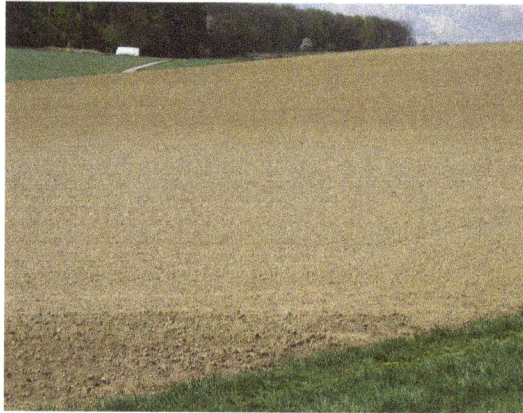
Loess field in Germany

Soil is vitally important to all life on land. It supports rooted plants, provides a habitat and shelter to many animals, and it is the home to bacteria, fungi, and other microorganisms that recycle organic material for reuse by plants.

While the general concept of soil is well established, the definition of soil varies, according to the perspective of the discipline or occupation using soil as a resource.

Soil is among our most important natural resources because of its position in the landscape and its dynamic, physical, chemical, and biologic functions. It has been both used and misused. On the positive side, human creativity is expressed in using soil for agriculture, gardening and landscaping, utilizing peat as an energy source, producing fertilizers to replenish lost nutrients, employing soils as building materials and transforming clay into eating and drinking vessels, storage containers, and works of art. On the other hand, anthropogenic activities have included fostering soil erosion and desertification through clear-cutting and overgrazing livestock, and contaminating soils by the dumping of industrial or household wastes.

The understanding of soil is incomplete. Despite the duration of humanity's dependence on and curiosity about soil, exploring the diversity and dynamic of this resource continues to yield fresh discoveries and insights. New avenues of soil research are compelled by our need to understand soil in the context of climate change and carbon sequestration. Our interest in maintaining the planet's biodiversity and in exploring past cultures has also stimulated renewed interest in achieving a more refined understanding of soil.

The earth's soil in general is sometimes referred to as comprising the pedosphere, which is positioned at the interface of the lithosphere with the biosphere, atmosphere, and hydrosphere. The scientific study of soil is called pedology or edaphology. Pedology is the study of soil in its natural setting, while edaphology is the study of soil in relation to soil-dependent uses.

Soil Components

Mineral Material

The majority of material in most soil is mineral. This consists of small grains broken off from the underlying rock or sometimes transported in from other areas by the action of water and wind. Larger mineral particles called *sand* and smaller particles called *silt* are the product of physical weathering, while even smaller particles called *clay* (a group of hydrous aluminium phyllosilicate minerals typically less than 2 micrometers in diameter) is generally the product of chemical weathering of silicate-bearing rocks. Clays are distinguished from other small particles present in soils such as silt by their small size, flake or layered shape, affinity for water and tendency toward high plasticity.

The mineral part of soil slowly releases nutrients that are needed by plants, such as potassium, calcium, and magnesium. Recently formed soil, for instance that formed from lava recently released from a volcano, is richer in nutrients and so is more fertile.

Organic Material

As plants and animals die and decay they return organic (carbon-bearing) material to the soil. Organic material tends to loosen up the soil and make it more productive for plant growth. Microorganisms, such as bacteria, fungi, and protists feed on the organic material and in the process release nutrients that can be reused by plants. The microorganisms themselves can form a significant part of the soil.

Water and Air

Soil almost always contains water and air in the spaces between the mineral and organic particles. Most soil organisms thrive best when the soil contains about equal volumes of water and air.

Soil Classification

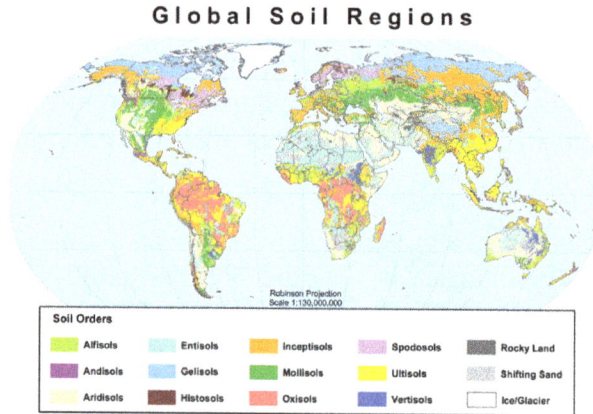

Map of global soil regions from the USDA

The World Reference Base for Soil Resources (WRB) is the international standard soil classification system. Development of this system was coordinated by the International Soil Reference and Information Centre (ISRIC) and sponsored by the International Union of Soil Sciences (IUSS) and the Food and Agriculture Organization (FAO) via its Land and Water Development division.

The WRB borrows from modern soil classification concepts, including United States Department of Agriculture (USDA) soil taxonomy. The classification is based mainly on soil morphology as an expression of pedogenesis, the creation of soil. A major difference with USDA soil taxonomy is that soil climate is not part of the system, except in so far as climate influences soil profile characteristics.

The WRB structure is either nominal, giving unique names to soils or landscapes, or descriptive, naming soils by their characteristics such as red, hot, fat, or sandy. Soils are distinguished by obvious characteristics, such as physical appearance (e.g., color, texture, landscape position), performance (e.g., production capability, flooding), and accompanying vegetation. A vernacular distinction familiar to many is classifying texture as heavy or light. *Light soils* have lower clay content than *heavy soils*. They often drain better and dry out sooner, giving them a lighter color. Lighter soils, with their lower moisture content and better structure, take less effort to turn and cultivate. Contrary to popular belief light soils do not weigh less than heavy soils on an air dry basis nor do they have more porosity.

Soil Characteristics

Soils tend to develop an individualistic pattern of horizontal zonation under the influence of site specific soil-forming factors. Soil color, soil structure, and soil texture are especially important components of soil morphology.

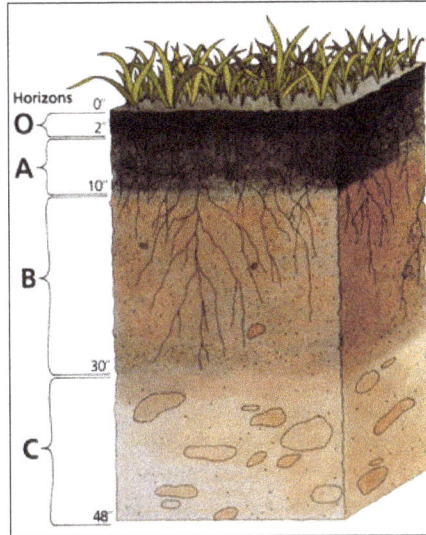

Soil horizons are formed by combined biological, chemical and physical alterations.

Soil color is the first impression one has when viewing soil. Striking colors and contrasting patterns are especially memorable. The Red River of the United States carries sediment eroded from extensive reddish soils like Port Silt Loam in Oklahoma. The Yellow River in China carries yellow sediment from eroding loessal soils. Mollisols in the Great Plains are darkened and enriched by organic matter. Podsols in boreal forests have highly contrasting layers due to acidity and leaching.

Soil color is primarily influenced by soil mineralogy. The extensive and various iron minerals in soil are responsible for an array of soil pigmentation. Color development and distribution of color within a soil profile result from chemical weathering, especially redox reactions. As the primary minerals in soil-parent material weather, the elements combine into new and colorful compounds. Iron forms secondary minerals with a yellow or red color; organic matter decomposes into black and brown compounds; and manganese forms black mineral deposits. These pigments give soil its various colors and patterns and are further affected by environmental factors. Aerobic conditions produce uniform or gradual color changes, while reducing environments result in disrupted color flow with complex, mottled patterns and points of color concentration.

Soil structure is the arrangement of soil particles into aggregates. These may have various shapes, sizes and degrees of development or expression. Soil structure influences aeration, water movement, erosion resistance, and root penetration. Observing structure gives clues to texture, chemical and mineralogical conditions, organic content, biological activity, and past use, or abuse.

Surface soil structure is the primary component of tilth. Where soil mineral particles are both separated and bridged by organic-matter-breakdown products and soil-biota exudates, it makes the soil easy to work. Cultivation, earthworms, frost action, and rodents mix the soil. This activity decreases the size of the peds to form a granular (or crumb) structure. This structure allows for good porosity and easy movement of air and water. The combination of ease in tillage, good moisture and air-handling capabilities, good structure for planting and germination are definitive of good tilth.

Soil texture refers to sand, silt and clay composition in combination with gravel and larger-material content. Clay content is particularly influential on soil behavior due to a high retention capacity for nutrients and water. Due to superior aggregation, clay soils resist wind and water erosion better than silty and sandy soils. In medium-textured soils, clay can tend to move downward through the soil profile to accumulate as illuvium in the subsoil. The lighter-textured, surface soils are more responsive to management inputs, but also more vulnerable to erosion and contamination.

Texture influences many physical aspects of soil behavior. Available water capacity increases with silt and, more importantly, clay content. Nutrient-retention capacity tends to follow the same relationship. Plant growth, and many uses which rely on soil, tends to favor medium-textured soils, such as loam and sandy loam. A balance in air and water-handling characteristics within medium-textured soils are largely responsible for this.

Water

Water is one of the Earth's naturally occuring element. Water is a substance composed of the chemical elements hydrogen and oxygen and existing in gaseous, liquid, and solid states. It is one of the most plentiful and essential of compounds. A tasteless and odourless liquid at room temperature, it has the important ability to dissolve many other substances. Indeed, the versatility of water as a solvent is essential to living organisms. Life is believed to have originated in the aqueous solutions of the world's oceans, and living organisms depend on aqueous solutions, such as blood and digestive juices, for biological processes. In small quantities water appears colourless, but water actually has an intrinsic blue colour caused by slight absorption of light at red wavelengths.

Although the molecules of water are simple in structure (H_2O), the physical and chemical properties of the compound are extraordinarily complicated, and they are not typical of most substances found on Earth. For example, although the sight of ice cubes floating in a glass of ice water is common place, such behaviour is unusual for chemical entities. For almost every other compound, the solid state is denser than the liquid state; thus, the solid would sink to the bottom of the liquid. The fact that ice floats on water is exceedingly important in the natural world, because the ice that forms on ponds and lakes in cold areas of the world acts as an insulating barrier that protects the aquatic life below. If ice were denser than liquid water, ice forming on a pond would sink, thereby exposing more water to the cold temperature. Thus, the pond would eventually freeze throughout, killing all the life-forms present.

Water occurs as a liquid on the surface of Earth under normal conditions, which makes it invaluable for transportation, for recreation, and as a habitat for a myriad of plants and animals. The fact that water is readily changed to a vapour (gas) allows it to be transported through the atmosphere from the oceans to inland areas where it condenses and, as rain, nourishes plant and animal life.

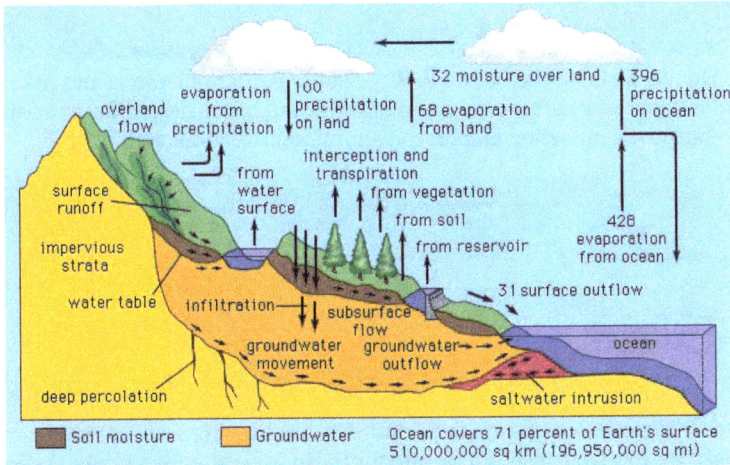

In the hydrologic cycle, water is transferred between the land surface, the ocean, and the atmosphere. The numbers on the arrows indicate relative water fluxes.

Because of its prominence, water has long played an important religious and philosophical role in human history. In the 6th century BCE, Thales of Miletus, sometimes credited for initiating Greek philosophy, regarded water as the sole fundamental building block of matter:

It is water that, in taking different forms, constitutes the earth, atmosphere, sky, mountains, gods and men, beasts and birds, grass and trees, and animals down to worms, flies and ants. All these are different forms of water. Meditate on water.

Two hundred years later, Aristotle considered water to be one of four fundamental elements, in addition to earth, air, and fire. The belief that water was a fundamental substance persisted for more than 2,000 years until experiments in the second half of the 18th century showed that water is a compound made up of the elements hydrogen and oxygen.

The water on the surface of Earth is found mainly in its oceans (97.25 percent) and polar ice caps and glaciers (2.05 percent), with the balance in freshwater lakes, rivers, and groundwater. As Earth's population grows and the demand for fresh water increases, water purification and recycling become increasingly important. Interestingly, the purity requirements of water for industrial use often exceed those for human consumption. For example, the water used in high-pressure boilers must be at least

99.999998 percent pure. Because seawater contains large quantities of dissolved salts, it must be desalinated for most uses, including human consumption.

The Hoover Dam on the Colorado River at the border of Nevada and Arizona demonstrates how natural resources of water can be harnessed for a variety of purposes, including human consumption, irrigation, and industry.

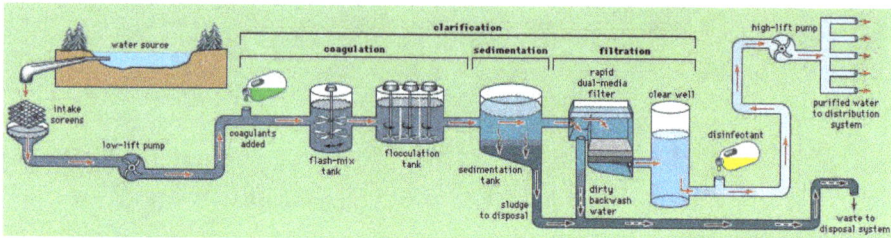

Water treatment systems are important for desalinating seawater so it can be used for human consumption and for purifying water for industrial use.

Structure of Water

Liquid Water

The water molecule is composed of two hydrogen atoms, each linked by a single chemical bond to an oxygen atom. Most hydrogen atoms have a nucleus consisting solely of a proton. Two isotopic forms, deuterium and tritium, in which the atomic nuclei also contain one and two neutrons, respectively, are found to a small degree in water. Deuterium oxide (D_2O), called heavy water, is important in chemical research and is also used as a neutron moderator in some nuclear reactors.

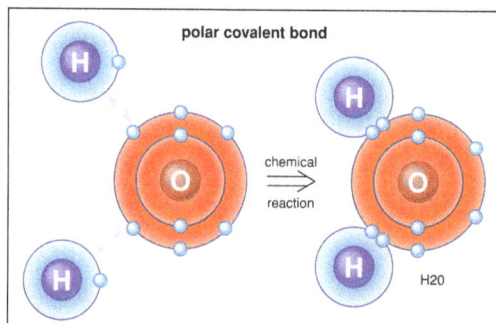

A water molecule is made up of two hydrogen atoms and one oxygen atom. A single oxygen atom contains six electrons in its outer shell, which can hold a total of eight electrons. When two hydrogen atoms are bound to an oxygen atom, the outer electron shell of oxygen is filled.

Although its formula (H_2O) seems simple, water exhibits very complex chemical and physical properties. For example, its melting point, o °C (32 °F), and boiling point, 100 °C (212 °F), are much higher than would be expected by comparison with analogous compounds, such as hydrogen sulfide and ammonia. In its solid form, ice, water is less dense than when it is liquid, another unusual property. The root of these anomalies lies in the electronic structure of the water molecule.

The water molecule is not linear but bent in a special way. The two hydrogen atoms are bound to the oxygen atom at an angle of 104.5°.

The O−H distance (bond length) is 95.7 picometres (9.57×10^{-11} metres, or 3.77×10^{-9} inches). Because an oxygen atom has a greater electronegativity than a hydrogen atom, the O−H bonds in the water molecule are polar with the oxygen bearing a partial negative charge (δ−) and the hydrogens having a partial positive charge (δ+).

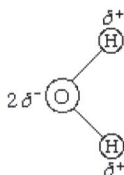

Hydrogen atoms in water molecules are attracted to regions of high electron density and can form weak linkages, called hydrogen bonds, with those regions. This means that the hydrogen atoms in one water molecule are attracted to the nonbonding electron pairs of the oxygen atom on an adjacent water molecule. The structure of liquid water is believed to consist of aggregates of water molecules that form and re-form continually. This short-range order, as it is called, accounts for other unusual properties of water, such as its high viscosity and surface tension.

Water droplets: Water is a polar molecule and is attracted to other polar molecules. Thus, droplets, or beads, of water form on a nonpolar surface because water molecules adhere together instead of adhering to the surface.

An oxygen atom has six electrons in its outer (valence) shell, which can hold a total of eight electrons. When an oxygen atom forms a single chemical bond, it shares one of its own electrons with the nucleus of another atom and receives in return a share of an electron from that atom. When bonded to two hydrogen atoms, the outer electron shell of the oxygen atom is filled.

The electron arrangement in the water molecule can be represented as follows:

Each pair of dots represents a pair of unshared electrons (i.e., the electrons reside on only the oxygen atom). This situation can also be depicted by placing the water molecule in a cube.

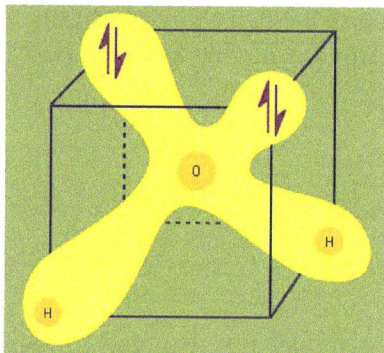

Each ↑↓ symbol represents a pair of unshared electrons. This electronic structure leads to hydrogen bonding.

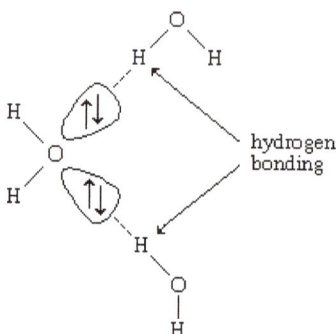

Structures of Ice

In the solid state (ice), intermolecular interactions lead to a highly ordered but loose structure in which each oxygen atom is surrounded by four hydrogen atoms; two of these hydrogen atoms are covalently bonded to the oxygen atom, and the two others (at longer distances) are hydrogen bonded to the oxygen atom's unshared electron pairs.

This open structure of ice causes its density to be less than that of the liquid state, in which the ordered structure is partially broken down and the water molecules are (on average) closer together. When water freezes, a variety of structures are possible depending on the conditions. Eighteen different forms of ice are known and can be interchanged by varying external pressure and temperature.

Significance of the Structure of Liquid Water

The liquid state of water has a very complex structure, which undoubtedly involves considerable association of the molecules. The extensive hydrogen bonding among the molecules in liquid water produces much larger values for properties such as viscosity, surface tension, and boiling point than are expected for a typical liquid containing small molecules. For example, based on the size of its molecules, water would be expected to have a boiling point nearly 200 °C (360 °F) lower than its observed boiling point. In contrast to the condensed states (solid and liquid) of water, which exhibit extensive association among the water molecules, its gaseous (vapour) phase contains relatively independent water molecules at large distances from each other.

The polarity of the water molecule plays a major part in the dissolution of ionic compounds during the formation of aqueous solutions. Earth's oceans contain vast amounts of dissolved salts, which provide a great natural resource. In addition, the hundreds of chemical reactions that occurs every instant to keep organisms alive all take place in aqueous fluids. Also, the ability of foods to be flavoured as they are cooked is made possible by the solubility in water of such substances as sugar and salt. Although the solubility of substances in water is an extremely complex process, the interaction between the polar water molecules and the solute (i.e., the substance being dissolved) plays a major role. When an ionic solid dissolves in water, the positive ends of the water molecules are attracted to the anions, while their negative ends are attracted to the cations. This process is called hydration. The hydration of its ions tends to cause a salt to break apart (dissolve) in the water. In the dissolving process the strong forces present between the positive and negative ions of the solid are replaced by strong water-ion interactions.

When ionic substances dissolve in water, they break apart into individual cations and anions. For instance, when sodium chloride (NaCl) dissolves in water, the resulting solution contains separated Na^+ and Cl^- ions.

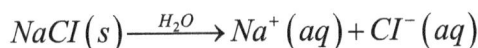

$$NaCl(s) \xrightarrow{H_2O} Na^+(aq) + Cl^-(aq)$$

In this equation the (s) represents the solid state, and the (aq), which is an abbreviation for aqueous, shows that the ions are hydrated—that is, they have a certain number of water molecules attached to them. As sodium chloride dissolves, four water molecules closely associate with the sodium ion. (The hydration number of Na^+ is four.) Just outside this inner hydration sphere is a region where water molecules are partially ordered by the presence of the $[Na(H_2O)_4]^+$ hydrated ion. This partially ordered region blends into "regular" (bulk) liquid water.

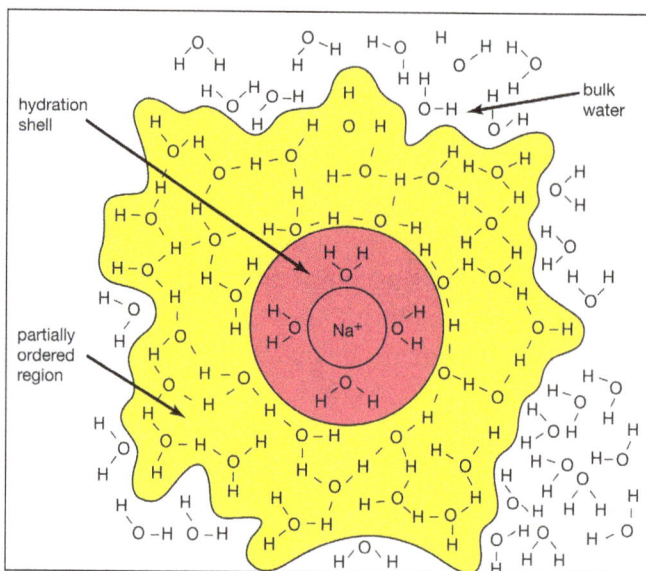

The hydration of a sodium ion

Generally speaking, the greater the charge density (the ratio of charge to surface area) of an ion the larger the hydration number will be. As a rule, negative ions have smaller hydration numbers than positive ions because of the greater crowding that occurs when the hydrogen atoms of the water molecules are oriented toward the anion.

Many nonionic compounds are also soluble in water. For example, ethanol (C_2H_5OH), the alcoholic component of wine, beer, and distilled spirits, is highly soluble in water. These beverages contain varying percentages of ethanol in aqueous solution with other substances. Ethanol is so soluble in water because of the structure of the alcohol molecule. The molecule contains a polar $O-H$ bond like those in water, which allows it to interact effectively with water.

There are many substances for which water is not an acceptable solvent. Animal fat, for example, is insoluble in pure water because the nonpolar nature of fat molecules renders them incompatible with polar water molecules. In general, polar and ionic substances are soluble in water. A useful rule of thumb for determining whether two substances are likely to be miscible (i.e., will mix to form a solution) is "like dissolves like." That is, two polar substances are likely to mix to form a solution, as are two nonpolar substances.

Behaviour and Properties

Water at High Temperatures and Pressures

The characteristic ability of water to behave as a polar solvent (dissolving medium) changes when water is subjected to high temperatures and pressures. As water becomes hotter, the molecules seem much more likely to interact with nonpolar molecules. For example, at 300 °C (572 °F) and high pressure, water has dissolving properties very similar to acetone (CH_3COCH_3), a common organic solvent.

Water exhibits particularly unusual behaviour beyond its critical temperature and pressure (374 °C [705.2 °F], 218 atmospheres). Above its critical temperature, the distinction between the liquid and gaseous states of water disappears—it becomes a supercritical fluid, the density of which can be varied from liquidlike to gaslike by varying its temperature and pressure. If the density of supercritical water is high enough, ionic solutes are readily soluble, as is true for "normal" water; but, surprisingly, this supercritical fluid can also readily dissolve nonpolar substances—something ordinary water cannot do. Because of its ability to dissolve nonpolar substances, supercritical water can be used as a combustion medium for destroying toxic wastes. For example, organic wastes can be mixed with oxygen in sufficiently dense supercritical water and combusted in the fluid; the flame actually burns "underwater." Oxidation in supercritical water can be used to destroy a wide variety of hazardous organic substances with the advantage that a supercritical-water reactor is a closed system, so there are no emissions released into the atmosphere.

Physical Properties

Water has several important physical properties. Although these properties are familiar because of the omnipresence of water, most of the physical properties of water are quite atypical. Given the low molar mass of its constituent molecules, water has unusually large values of viscosity, surface tension, heat of vaporization, and entropy of vaporization, all of which can be ascribed to the extensive hydrogen bonding interactions present in liquid water. The open structure of ice that allows for maximum hydrogen bonding explains why solid water is less dense than liquid water—a highly unusual situation among common substances.

Selected physical properties of water	
Molar mass	18.0151 grams per mole
Melting point	0.00 °c
Boiling point	100.00 °c
Maximum density (at 3.98 °C)	1.0000 grams per cubic centimetre
Density (25 °C)	0.99701 grams per cubic centimetre
Vapour pressure (25 °C)	23.75 torr

Heat of fusion (0 °C)	6.010 kilojoules per mole
Heat of vaporization (100 °C)	40.65 kilojoules per mole
Heat of formation (25 °C)	−285.85 kilojoules per mole
Entropy of vaporization (25 °C)	118.8 joules per °C mole
Viscosity	0.8903 centipoise
Surface tension (25 °C)	71.97 dynes per centimeter

Chemical Properties

Acid-base Reactions

Water undergoes various types of chemical reactions. One of the most important chemical properties of water is its ability to behave as both an acid (a proton donor) and a base (a proton acceptor), the characteristic property of amphoteric substances. This behaviour is most clearly seen in the autoionization of water:

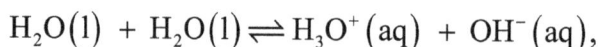

$$H_2O(l) + H_2O(l) \rightleftharpoons H_3O^+(aq) + OH^-(aq),$$

where, the (l) represents the liquid state, the (aq) indicates that the species are dissolved in water, and the double arrows indicate that the reaction can occur in either direction and an equilibrium condition exists. At 25 °C (77 °F) the concentration of hydrated H^+ (i.e., H_{3O}^+, known as the hydronium ion) in water is 1.0×10^{-7} M, where M represents moles per litre. Since one OH^- ion is produced for each H_3O^+ ion, the concentration of OH^- at 25 °C is also 1.0×10^{-7} M. In water at 25 °C the H_3O^+ concentration and the OH^- concentration must always be 1.0×10^{-14}:

$$\left[H^+\right]\left[OH^-\right] = 1.0 \times 10^{-14},$$

where, $[H^+]$ represents the concentration of hydrated H^+ ions in moles per litre and $[OH^-]$ represents the concentration of OH^- ions in moles per litre.

When an acid (a substance that can produce H^+ ions) is dissolved in water, both the acid and the water contribute H^+ ions to the solution. This leads to a situation in which the H^+ concentration is greater than 1.0×10^{-7} M. Since it must always be true that $[H^+]$ $[OH^-] = 1.0 \times 10^{-14}$ at 25 °C, the $[OH^-]$ must be lowered to some value below 1.0×10^{-7}. The mechanism for reducing the concentration of OH^- involves the reaction,

$$H^+ + OH^- \rightarrow H_2O,$$

which occurs to the extent needed to restore the product of $[H^+]$ and $[OH^-]$ to 1.0×10^{-14} M. Thus, when an acid is added to water, the resulting solution contains more H^+ than OH^-; that is, $[H^+] > [OH^-]$. Such a solution (in which $[H^+] > [OH^-]$) is said to be acidic.

The most common method for specifying the acidity of a solution is its pH, which is defined in terms of the hydrogen ion concentration:

$$pH = -\log\left[H^+\right],$$

where the symbol log stands for a base-10 logarithm. In pure water, in which $[H^+] = 1.0 \times 10^{-7}$ M, the pH = 7.0. For an acidic solution, the pH is less than 7. When a base (a substance that behaves as a proton acceptor) is dissolved in water, the H^+ concentration is decreased so that $[OH^-] > [H^+]$. A basic solution is characterized by having a pH > 7. in aqueous solutions at 25 °C:

Neutral solution	$[H^+] = [OH^-]$	pH = 7
Acidic solution	$[H^+] > [OH^-]$	pH < 7
Basic solution	$[OH^-] > [H^+]$	pH > 7

Oxidation-Reduction Reactions

When an active metal such as sodium is placed in contact with liquid water, a violent exothermic (heat-producing) reaction occurs that releases flaming hydrogen gas.

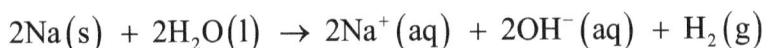

$$2Na(s) + 2H_2O(l) \rightarrow 2Na^+(aq) + 2OH^-(aq) + H_2(g)$$

This is an example of an oxidation-reduction reaction, which is a reaction in which electrons are transferred from one atom to another. In this case, electrons are transferred from sodium atoms (forming Na^+ ions) to water molecules to produce hydrogen gas and OH^- ions. The other alkali metals give similar reactions with water. Less-active metals react slowly with water. For example, iron reacts at a negligible rate with liquid water but reacts much more rapidly with superheated steam to form iron oxide and hydrogen gas.

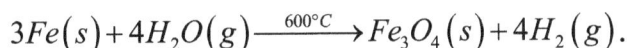

$$3Fe(s) + 4H_2O(g) \xrightarrow{600°C} Fe_3O_4(s) + 4H_2(g).$$

References

- Mineral-chemical-compound, science: britannica.com, Retrieved 10 August, 2019

- Classification-of-minerals, mineral-chemical-compound, science: britannica.com, Retrieved 11 March, 2019

- Rock-geology, science: britannica.com, Retrieved 18 May, 2019

- Types-of-rock: zmescience.com, Retrieved 17 January, 2019

- Soil: newworldencyclopedia.org, Retrieved 7 April, 2019

- Water, science: britannica.com, Retrieved 13 February, 2019

Permissions

All chapters in this book are published with permission under the Creative Commons Attribution Share Alike License or equivalent. Every chapter published in this book has been scrutinized by our experts. Their significance has been extensively debated. The topics covered herein carry significant information for a comprehensive understanding. They may even be implemented as practical applications or may be referred to as a beginning point for further studies.

We would like to thank the editorial team for lending their expertise to make the book truly unique. They have played a crucial role in the development of this book. Without their invaluable contributions this book wouldn't have been possible. They have made vital efforts to compile up to date information on the varied aspects of this subject to make this book a valuable addition to the collection of many professionals and students.

This book was conceptualized with the vision of imparting up-to-date and integrated information in this field. To ensure the same, a matchless editorial board was set up. Every individual on the board went through rigorous rounds of assessment to prove their worth. After which they invested a large part of their time researching and compiling the most relevant data for our readers.

The editorial board has been involved in producing this book since its inception. They have spent rigorous hours researching and exploring the diverse topics which have resulted in the successful publishing of this book. They have passed on their knowledge of decades through this book. To expedite this challenging task, the publisher supported the team at every step. A small team of assistant editors was also appointed to further simplify the editing procedure and attain best results for the readers.

Apart from the editorial board, the designing team has also invested a significant amount of their time in understanding the subject and creating the most relevant covers. They scrutinized every image to scout for the most suitable representation of the subject and create an appropriate cover for the book.

The publishing team has been an ardent support to the editorial, designing and production team. Their endless efforts to recruit the best for this project, has resulted in the accomplishment of this book. They are a veteran in the field of academics and their pool of knowledge is as vast as their experience in printing. Their expertise and guidance has proved useful at every step. Their uncompromising quality standards have made this book an exceptional effort. Their encouragement from time to time has been an inspiration for everyone.

The publisher and the editorial board hope that this book will prove to be a valuable piece of knowledge for students, practitioners and scholars across the globe.

Index

www.ingramcontent.com/pod-product-compliance
Lightning Source LLC
Chambersburg PA
CBHW061938190326
41458CB00009B/2769

*9 7 8 1 6 4 1 7 2 4 6 8 5 *